GREEN IMMIGRANTS

The Plants That Transformed America

Claire Shaver Haughton

Drawings by Russell Peterson

HARCOURT BRACE JOVANOVICH

GREEN IMMI- GRANTS

A Harvest/HBJ Book

NEW YORK AND LONDON

Printed in the United States of America

LIBRARY OF CONGRESS CATALOGING IN PUBLICATION DATA
Haughton, Claire Shaver.
Green immigrants.
(A Harvest/HBJ Book)
Includes index.
1. Plant introduction—United States—History—Dictionaries.
2. Plant introduction—History—Dictionaries.
3. Plants, Cultivated—United States—History—Dictionaries.
4. Plants, Cultivated—History—Dictionaries.
I. Title.
[SB108.U5H68 1980]   631′.0973   79-24258
ISBN 0-15-636492-1

First Harvest/HBJ edition 1980
ABCDEFGHIJ

CONTENTS

vii ❧ CONTENTS ❧

S

T

U

V

W

X   Y   Z

PREFACE

One of the disconcerting facts brought to light by World War II was that the average G.I. knew little about his country's history. I participated in one earnest project resulting from that discovery, the preparation of a series of movies in American history designed for the public schools. I was assigned the early colonial years, and promptly confronted a basic question: What did those first colonists eat?

There were no gardens or farms in America. A great arc of forests covered much of the Atlantic coast. What seeds did the colonists bring with them? Why did they consider them essential? How well did plants from the Old World thrive in the New?

I found there was no book that told this story. Yet immigrant plants, from those that grew in the Pilgrims' herb gardens to the strange new fruits and vegetables from Puerto Rico and other Caribbean islands that crowd our urban markets today, have played a part in our nation's history, influencing it as surely as the diverse people who settled here.

The lack of information about immigrant plants was a challenge, and my research became absorbing. Haunting libraries, botanical gardens, agricultural centers, and universities, I discovered that our gardens and farms, our roadsides and fields, have been transformed by foreign plants. Most of our cultivated flowers, vegetables, and fruits, of our grains, grasses, and clovers, and nearly 70% of our weeds, have come to the United States from other nations. Botany constitutes an important, if neglected, chapter in American history.

Green Immigrants investigates that chapter. It relates the history and romance, the legend and folklore, of nearly one hundred growing plants, telling where they came from, how they arrived here, and what has happened to them since.

The green invasion began with Columbus. The Spanish immediately discovered the fertility of this new world and lost no time in establishing great plantations of European fruits and grains for export to the Old World. A century of fabulous success in colonizing Spanish America was largely due to the padres, who were trained in horticulture as well as in religion. They not only supervised the planting of mission fields and haciendas with European grains, vegetables, and fruits, but also collected the plants and flowers of America and shipped them to Spain, where they were guarded as jealously as the gold and silver of the conquistadors. By one of the ironies of history, many of them, including the potato and tomato of the Andes, were shipped to the United States centuries later as European products.

At the beginning of the seventeenth century, Catholic settlers under Champlain established French Canada on the

Saint Lawrence River, and Dutch Protestants founded New Netherland on an island at the mouth of the Hudson. At about the same time the English were founding the Jamestown colony in Virginia, and Captain John Smith issued his stirring call:

> Here is land enough for all. Rich soil and freedom and adventure;—there is nothing in America that men with strong hands and brave hearts cannot possess. Sail then! Sail westward—to share in the building of this great country of the west!

The Pilgrims arrived in New England in 1620, the Swedes and other Scandinavians on the Delaware not much more than a decade later. All Europe was in a state of religious and social unrest, and every nation was eager to be rid of its malcontents. Taking a cue from John Smith, England opened her colonies to men of all nations and religions.

They came by the hundreds and thousands—the Quakers, the Catholics, the Huguenots, and the various religious sects of Germany whom we know as the Pennsylvania Dutch. Their dreams and their seeds came with them, for they were all devoted horticulturists. Immigrant cotton, sugar, rice, indigo, and native tobacco built the great plantations of our south. The Puritans founded a secure economy in New England. The people of New Netherland became New Yorkers and, with the Pennsylvania Dutch and the Huguenots, created rich farms and shipping centers for the middle colonies.

The democracy in England's policies welded these diverse peoples into the unity of thirteen colonies which became the foundation of our republic. Gifted leaders with creative ideas came with the settlers and opened new vistas of economic wealth. Later generations developed agricultural machinery, and hybridized more productive and nutritious plants to provide abundant food, fibers, and plastics for an expanding economy.

The role green immigrants have played in building our American democracy is summed up by Charles Morrow Wilson, a twentieth-century writer, who declared that the essence of American freedom is epitomized in the introduction of new plants and man's freedom to own and plant the land as he pleases.

GREEN IMMIGRANTS

A IS FOR APPLE

Apples came to Britain with the Roman conquest, some 2,000 years ago, and quite possibly Julius Caesar himself ordered that apple trees be planted. He took a keen interest in botany. An herbalist traveled with his legions, and a great collection of foreign plants grew in his garden on the Hill of Janus. This garden was one of the loveliest in Rome, and he willed it to the people.

Slowly, during the hundreds of years that Rome ruled Britain, the country was transformed into a typical Roman province. While the legions built roads that radiated from London, and towns through which their civilization was disseminated, the officers built villas with walled gardens and they planted apple orchards. Apple trees sprang up about the native villages, too, but the Romans laughed at this evidence that Britons had plundered an orchard here and there, and called the apple the "stamp of Rome."

The new religion, Christianity, which came to Roman Britain fairly early, had its roots in that older religion whose history began with the tale of a man and a woman in a garden—and an apple. As the apple became a symbol of the Roman occupation, monasteries became the symbol of Christianity, which had supplanted the Druid faith of the Britons. And most monasteries had apple orchards.

Rome withdrew from Britain by the middle of the fifth century, whereupon Germanic tribes—the Anglo-Saxons and the Jutes—began a series of raids that developed into large-scale invasions, and then into settlement. The Roman villas, gardens, and orchards were abandoned, but the three great gifts of Rome—law, Christianity, and the apple—survived. Wherever monasteries or convents were established, orchards were established, too, until in the end this fruit that had arrived with the Romans became a British staple.

But apples are not Roman. They are thought to be indigenous to the Caucasus Mountains of western Asia, and perhaps to Anatolia, the Asian part of Turkey. They have been cultivated since prehistoric times—longer, it may well be, than any other fruit—as archeological evidence clearly shows.

Apples in the New World

The Spanish, long the leaders in colonizing the New World, followed the example of the Romans in Britain, and introduced their law, their Catholic faith, and many of their plants. But apple trees need 900 to 1,000 hours of temperatures below 45° F. to store up the energy that will explode their dormant buds in spring; and the Spanish soon discovered that they would not thrive in the subtropics. So it was left to the Puritans to introduce apples to America.

The story began early in the seventeenth century, at

Cambridge University, where a young man named William Blackstone—no forebear of the eminent eighteenth-century jurist and author of the *Commentaries*—was studying for ordination in the Church of England. Botany and agronomy were fashionable subjects at both Oxford and Cambridge, and it was incumbent upon young gentlemen of property to be informed about them. After taking his degree in theology, Blackstone hurried home to try his hand at apple culture, and in a few years announced the development of a new variety, the yellow sweeting.

The Puritanism sweeping England had already led to the establishment of the colony at Plymouth, in Massachusetts. Members of the government and of the established church, disturbed because the Puritans already had a strong foothold in New England, organized a group to found an Episcopal colony there, too. The guiding spirit in this enterprise was Ferdinando Gorges, whose ambition it was to make all New England a royal colony and to preside over it as governor.

William Blackstone was chosen to be minister to these settlers, who arrived in Boston Harbor in 1623. He came well equipped, with his Bible, many boxes of books, and a bag of apple seeds, and immediately fell in love with the wilderness. Beacon Hill, which rose 100 feet above the harbor and had been designated by Gorges as the place for beacon fires in case of Indian attack, would be, Blackstone decided, an ideal site for his new orchard. Proceeding on the principle of brotherly love, he persuaded the Indians to clear the hill, plant his apple seeds, and help him build a tight cabin, in which he installed his library.

Meanwhile, the colonists built their cabins along the shore. Unlike Blackstone, they found the savages, the wilderness, and the bitter winter not at all to their taste. They starved and quarreled, and when spring came, bringing with it swarms of mosquitoes from the swamps, they sailed back to England. But Blackstone's apple seeds were sprouting, and he refused to leave. There on Beacon Hill he remained for seven years, alone among the Indians with his books and his apple trees.

When John Winthrop and a new band of Puritans sailed into Boston Harbor in 1630, in seventeen ships, Blackstone's trees were bearing their first apples. It was a meager crop, but he shared it with the new colonists, and recommended that they settle, as he had, on the high fertile ground of Beacon Hill. Unmoved by Blackstone's generosity, the Puritans decreed that an Episcopalian in their midst could not be tolerated; and so he departed to what is now Rhode Island. He carried with him another bag of apple seeds, and above the river that today bears his name he planted another orchard.

The Puritans, too, had brought apple seeds from England, and they, too, promptly planted orchards. In time their trees bloomed, but like those on Beacon Hill, they bore few apples. The town fathers put their heads together, recalled that English orchards had beehives scattered through them, and decided that American bumblebees were not interested in pollinating English apple blossoms. The next spring numerous hives of bees were shipped to Boston, and from then on the trees bore abundantly.

Governor Winthrop, who was an old Cambridge man himself, eventually sent Blackstone's library to him, along with a priceless hive of bees. With their aid Rhode Island orchards flourished, and Blackstone gave seeds and grafts to Roger Williams at Providence and to the adjacent Indian villages. Records show that some of his trees were still alive in 1837.

Apples soon grew throughout New England. In 1641 Governor Endicott of Salem purchased 250 acres of land on which 500 apple saplings were growing. Peregrine White, the first child born to English parents in New England—an event that occurred aboard the *Mayflower*, anchored in Cape Cod Bay, November 20, 1620—celebrated his coming of age by planting an apple orchard on his estate in Marshfield, south of Boston. By 1646 Massachusetts had passed her first law on "proper punishment" for anyone robbing an apple orchard, and by 1650 every Puritan gentleman of standing was trading in apple trees.

This was the beginning of apples in America.

Apples in a Growing Nation

Not only New England boasted apples. The foothills of the Appalachians provided an ideal environment for them, and so did the Virginia Tidewater, where gentlemen farmers often kept their plantation records in Latin. One of them was William Byrd of Westover, who reported in his "Secret Diary" that he was growing twenty-five varieties of apples, seven of which could be preserved all year. As to preservation, the colonists followed the English method, and packed apples in layers of dry moss in earthen pots that they sealed with resin and stored in cellars.

All the first apples in America were grown from seeds, which were easy to import and trade. But it soon became obvious that the results were undependable, because trees grown from the imported seed often crossed with wild native crab apples. Such natural crosses produced fruit with strange forms and flavors, often quite worthless. But some of these apples were superior and destined for lasting fame—among them the Baldwin of Massachusetts and the Northern Spy of New York. Which brings us to a settler in Ontario named John McIntosh, who took a fancy to a young apple tree he found one day growing wild beside the Saint Lawrence River. He dug it up, carried it home, and planted it in his garden, where it produced apples of notable excellence. Their fame spread, and a demand for them sprang up and increased. From that one tree, ultimately, countless grafts and seedlings were taken; McIntosh orchards proliferated, and their apples became known across the continent. The tree itself lived until

1908, and a monument now stands where it grew, in honor of the fame and fortune it brought to Ontario.

Apple culture grew with our nation. No farm was self-sustaining without its orchard of various kinds of apple. There were summer and winter apples, cooking and eating apples, and cider apples. A farm woman invented a great American dessert when she filled a crust with what she had most of, apples, and flavored them with maple sugar and a pinch of spice. Another confection was apple butter. Into a huge iron kettle over an open fire went bushel after bushel of peeled, sliced surplus apples that would not keep. Gallons of cider were added, a sugar loaf, and a grated nutmeg. All day someone had to stir this pungent mass, until it cooked down enough to be stored in gallon crockery pots.

Apple bark, which produced an excellent fast yellow color, was used throughout the colonies for dyeing wool and linen; isolated communities in the Appalachians did so until the present century.

In Pennsylvania Dutch country drying solved the problem of preserving apples. Untold bushels were peeled and sliced each year and dried in the sun or an oven; tied in great cotton bags, they hung from attic rafters or in chimney corners, waiting for the day they would be made into pie or *schnitz un knept*, apples with dumplings. For more than fifty years dried apples constituted one of the principal items shipped from Philadelphia to the West Indies.

These were the days of the cider press. According to statistics in an old book, about one farmer in ten had a cider press, and the average farm cellar stored from twenty to fifty barrels of cider. It was the hospitality drink for social and business callers, served free to passing travelers, peddlers, and Indians. And cider vinegar was essential in pickling everything from cucumbers to pigs' feet.

These were the days of the apple bin that made the winter evenings on isolated farms something to remember.

> And, for the winter fireside meet,
> Between the andirons' straddling feet,
> The mug of cider simmered slow,
> The apples sputtered in a row. . . .

So Whittier pictured, in "Snowbound," a nostalgic boyhood memory.

These were the days when farmers gathered at the crossroads store and boasted about the number and variety of apples they grew, each man considering himself an expert on budding, grafting, and cultivation.

These were the days when every covered wagon carried its bag of apple seeds out west. These, in fact, were the days of Johnny Appleseed, who gathered seeds from cider mills, hiked through the western wilderness, and shared them with pioneers and Indians, or planted them along streams and trails where hungry pioneers might pass.

Apples were carried along the Oregon Trail to the Pacific Northwest, where settlers developed their own type of root grafting and created varieties that have made it one of the great apple areas of the world.

Apples were taken to South America, and an astonishing note on their culture appears in Darwin's *Voyage of the Beagle*. The ship stopped at Chiloe, an island off the west coast of Chile, where "the inhabitants possess a marvelously short method of making an apple orchard. A branch as thick as a man's thigh is cut off at the joint, all small branches lopped off, and it is placed about two feet deep in the damp ground. During the summer this stump throws out long shoots and sometimes bears fruit. By the third summer, the stump is changed to a well-wooded tree, loaded with fruit." This, Darwin adds, he has seen for himself.

In general, climate limits the production of apples in South America. But in Australia orchards were producing satisfactorily fifty years after Sidney was founded.

Modern Apples

Apple seeds, which played such a dramatic part in the migration of orchards, are no longer sold by American nurseries, although as late as 1880 the Atlee Burpee Company offered a quart of them for ten dollars.

Pyrus malus, the ancient apple of the Near East, and *Pyrus baccata,* the crab apple, crossed to produce the ancestors of all the more than 200 varieties of apples that are now cultivated in the United States. They have been hybridized for qualities that make them easy to ship and store, for size and color appeal that will stimulate quick sales, and for a high yield. Today the same amount of acreage produces twice as many apples as it did in 1910.

The chief apple-growing regions of the world are in Western Europe, especially in France; but in North America, too, apple production is enormously high. Washington leads all the apple states; in the east, the Piedmont is a close rival to New England; and crops are grown in other areas as well, mostly for local consumption. It is estimated that, all told, there are 71 million trees in our commercial orchards, and that they yield 150 million bushels annually.

The most pressing problems of large commercial orchards are disease, birds, and labor costs. The U.S. Department of Agriculture and our agricultural colleges who carry on constant research into safe sprays and disease-resistant apples, have devised ingenious methods for discouraging birds, and have tested a variety of automatic devices for picking and packing apples. The University of Michigan has even sought to promote apple sales through health experiments. Four hundred students were persuaded to test the old jingle by eating an apple a day. For two terms they followed this regimen, and they did indeed prove healthier than other students: they had fewer colds and made "significantly fewer" trips to the college dispensary.

But in this Brave New World of mechanization, production control, and genetics, who can remember the incomparable taste of a wild apple? A walk down the lane, across the wood lot and the pasture, to pick a wild "coonie" or "pippin"? You shake it and hear the ripe seeds rattle—and even before your first bite you know the crisp tart sweetness of a wild apple eaten in the sharp October air.

AFRICAN VIOLET

When Germany set out in the 1880's to become an imperial power, the young and ambitious Emperor Wilhelm II appointed, as governor of German East Africa, one Walter von St. Paul-Illaire. Also young, and full of enthusiasm, St. Paul-Illaire decided to explore his vast domain, which extended from the Indian Ocean westward across jungle-covered plains to Lake Tanganyika, Lake Victoria, and the peaks of the Usambara Mountains.

As he and his party, led by Bantu and Usambara guides, climbed the foothills and the ragged unknown peaks, St. Paul-Illaire became more and more fascinated with the flora. One plant particularly captivated him. It grew in the crevices of rocks or fallen trees, and in pockets of rich earth, wherever the sun filtered through the thick tropic growth. The violet-like flowers, in delicate shades of blue, pink, and purple, were surrounded by ornamental leaves and looked like small corsages strewn by nature through the forest, just waiting to be gathered.

St. Paul-Illaire was no stranger to gardening—his father, Baron Walter von St. Paul-Illaire, was renowned as a horticulturist—but he had never seen anything like these plants. Under his direction an assortment was gathered and carefully packed, and when the expedition returned to Zanzibar he shipped them off to his father in Berlin. Quite unaware, he had launched a tropical beauty that would conquer the world.

Fascinated by the unknown African flower, the baron sent samples to the Royal Botanical Garden at Herrenhausen to be identified. The director, Herman Wentland, decided that they belonged to a genus previously unknown to botanists, and honored the baron and his son by naming it "Saintpaulia," with "Violette Usambara" as its common name. He found that there were two distinct species, one of which he called *S. diplotricha*. From the other, *S. ionantha*—a Greek word meaning violetlike—most of the varieties well known today have been derived.

At first the Saintpaulias were confined to botanical gardens and conservatories, because it was assumed that they would thrive only in conditions that simulated the hot and humid atmosphere of their tropic home. About 1900 a few specimens were sent to the conservatory at Cambridge University, and the flower-loving English soon took these new arrivals to their hearts, and began calling them "African violets." The popular name seemed entirely apt, but despite their appearance, the Saintpaulias are not violets at all. They belong to the Gesneriaceae family, and are close cousins to the gloxinias that grow in the rain forests of tropical Brazil.

From Conservatory to House Plant

England encouraged plant hunters to seek out new Saintpaulias in the equatorial forests of the Usambara Mountains, and the search was given added impetus, after World War I, when Germany lost all her African holdings and Tanganyika became British. New varieties of *S. ionantha* were found in the lower levels of the mountain range, and of *S. diplotricha* on the wooded heights. Toward the summit of Mount Tonque, which reaches 6,000 feet, a third species, *S. tonquenensis*, was discovered.

The spreading interest in African violets had encouraged

many possessors of green thumbs to experiment with them as house plants. To their astonished delight, they soon found them notably sturdy and adaptable, and infinitely varied as well. By 1925 African violets were being widely grown as house plants on the Continent as well as in England.

African Violets in the United States

Such popularity abroad encouraged the California firm of Armacot & Royster to test the commercial value of African violets in the United States. They sent to Europe for some seeds in 1926, and received a good assortment of the varieties that had proved most popular there: Blue Boy, Neptune, Admiral, Amethyst, Commodore. Perhaps their excellent performance was the springboard for the unparalleled popularity of the African violet in the United States.

Armacot & Royster were soon completely absorbed in African-violet culture. The fantastic changeability, the mutations, and the inherited variations of shape and size and color in both flowers and foliage convinced them that the Saintpaulia would challenge anyone who grew it. They saw in it a perfect collector's plant and an excellent commercial venture, but even so they were unprepared for the spontaneous demand that followed their first public offering. The African violet took the United States by storm. Today, less than half a century since it was introduced, it is the most widely grown house plant in the nation. Perhaps as many as a thousand recognized varieties exist today, and new ones continue to be developed by enthusiastic amateurs as well as by professional growers.

Variations and Their Causes

Many reasons are offered for the limitless variation of African violets. Some authorities believe that the mixed inheritance of many varieties has resulted in a change in the number or arrangement of the chromosomes in the cell nuclei. Others sug-

gest that the change from a stable, humid, tropical environment to the extremes of artificial light, heat, and humidity in a
modern house, to which the plants have been forced to adapt,
has caused the variations. Possibly both factors play a part,
combined with a basic tendency to mutation that manifests
itself in the rearrangement or increase in the number of petals,
and in the color, shape, size, and texture of the leaves.

The blooming habits of African violets vary, too. Many
bear flowers in four to nine months, and some continue
flowering for a year. The modern miniatures, and the newer
varieties with larger blooms—the Amazons, the Supremes, the
massive Duponts—usually bear fewer flowers than the old-
timers, but these are double and last longer.

Perhaps the greatest variation comes in the reproductive
patterns. Amateurs frequently propagate with leaf cuttings.
The new plantlets most frequently start at the base of a leaf
cutting, but they may appear on the stalk or the veins of the
leaf; and the leaves of different plants may produce at different stages of their maturity. Raising African violets from seed
is an even more interesting project, because you cannot predict what you will get, especially when crossing two varieties.
There is nothing monotonous about growing African violets.

Basic Requirements

Bearing in mind that these plants originated in the tropical
forests of Tanganyika's mountains should help us to remember that they need protection from extremes of cold, and
from direct sunlight. Another of their characteristics is a small
root system, developed through eons of growth in pockets of
earth in rocky terrain. They grow best in small containers,
which to the modern householder is another asset: one can
keep a wide assortment in a limited space.

Perhaps the beauty and adaptability of the African violet
have helped to make electric terrariums one of the fastest-
spreading fads in America. They have become the chief outlet

for plant lovers shut up in high-rise apartments or window-less, air-conditioned offices and stores; and they provide a pseudo-tropical environment in which African violets luxuriate. In the midst of twentieth-century mechanization they disarm their devotees, and growing them and other members of the Gesneriaceae family has become a widespread modern cult.

The National African Violet Society

The National African Violet Society, organized in 1946, fills a real and growing need. To help the thousands of amateurs lost among the violet's endless varieties, its registration committee eliminates duplication of names and establishes official nomenclature. The society also sets standards of selection, provides an opportunity for growers to share information, and assists experienced growers to exhibit their accomplishments in local, state, and national shows.

The society is, above all, a reliable source of information on the hybridizing, cultivation, and care of African violets, which seem far more familiar under this name than as Saintpaulias to window-sill gardeners and to the newspaper and magazine columns that cater to them. Indeed, the name is so well established today that most florists, nurserymen, and even botanists use it, and some amateurs are defensive about it.

It may be appropriate that the democratic public refuses to accept the name of Baron Walter von St. Paul-Illaire for its beloved flower. Gone and forgotten are the European powers and princes who grasped Africa's wealth less than a century ago, but Africa's wildflower has won international recognition and admiration.

The African violet possesses the charm of the unpredictable, the appeal of gambling. Who can resist it?

AMARANTH

The amaranths are old-fashioned flowers familiar to many gardeners as globe amaranths, love-lies-bleeding, cockscomb, Joseph's-coat, and prince's-feather, homely names by which they were known to the early settlers in America. Their luxurious foliage in shades of red, cream, gold, and green, their spikes and plumes and sprays of flowers were admired and coveted by gardeners down the years and through Victorian days. Though today they are seldom used, there is every indication that the twentieth century will yet see their return, not only to the flower garden as hybrid ornamentals, but as food on our grocers' shelves, because various amaranths are being investigated to help feed a hungry world.

History of the Ornamentals

Amaranths of one species or another appear to be indigenous to the Americas, Africa, southern Asia, and Australia. According to *Selected Weeds of the United States*, a Department of Agriculture handbook, the only one strictly native to the United States is *Amaranthus blitoides*, or prostrate pigweed. The ornamentals had to make their way from Asia into Europe before they could set out for American gardens.

The name "amaranth" comes from two Greek words, and means "not withering." The brightly colored bracts that surround the flowers of many species retain their freshness for an exceptional length of time after they are gathered. For this reason they became the symbol of immortality in ancient Greece, and wound their way through eons of Greek myths as "immortals." The early Christian church, too, adopted them as symbolic of immortality, and they found their place in monastery gardens during the Middle Ages. Their rich purples, royal reds, and golds blended perfectly with the exquisite colors of stained-glass windows, saints' robes, and altar hangings in church and shrine. The globe amaranth, in particular, is still much cultivated in Catholic countries, and continues to play a part in the religious ceremonies and church decoration of rural Spain and Portugal.

By Elizabethan days the amaranths had spread across Europe and were thriving in England, where they moved into the garden plots of cottage and castle. There are many records of their popularity in English literature. John Gerard, botanist and barber-surgeon, extravagantly praised Joseph's-coat, with its red, yellow, and green leaves, in *The Herball, or General Historie of Plantes*, published in 1597, saying in conclusion, "It farre exceedeth my skill to describe." John Rea, in his comprehensive garden book *Flora: Seu, De Florum Cultura*, published in 1665, wrote that "country women have named the exotic Floramor, Love-lies-a-bleeding, and consider it an old-fashioned flower." And in *Paradise Lost*, Milton crowns the angels with the immortal amaranth.

Medicinal properties were also ascribed to them. By the seventeenth century they were considered an essential herb, and the first English colonists carried their seeds to the New World.

Amaranths in America

Old Virginia records show that amaranths—or flower-gentles, as they were commonly called—were planted at Jamestown and Williamsburg, and in the walled gardens of many plantations and town houses in Virginia and the Carolinas. The summer gardens of the Governor's Palace at Williamsburg boasted exhibits of pink globe amaranth, carmine cockscomb, and magenta plumes of prince's-feather, waving their varied brilliance above the Queen Anne's lace. John Custis and William Byrd, Jefferson and Washington, grew varied amaranths in their Virginia gardens. They were dried and used with pearly everlasting and strawflowers in winter bouquets: arranged in choice glass and china basins filled with sand in lieu of flower holders, they graced the winter drawing rooms of many southern homes.

The Pennsylvania Dutch, who were incomparable gardeners, grew various amaranths with pride, and they flourished even in the gardens of Penn's Quakers. Perhaps this provision for pleasure seemed justified by the ready sale they found in the farmers' markets—markets that still thrive in Pennsylvania, and still sell amaranths for winter bouquets.

Early records show that love-lies-bleeding and globe amaranths were planted on Cape Cod and in the Bay Colony by the first settlers, who cherished them in their dooryards, growing them beside tansy, yarrow, and teasel. One of the first American nursery catalogues advertised the amaranth as an "effective plant with plumed tufts of lustrous crimson."

During the Victorian era they were much used as bedding plants in parks and on estates. But their popularity faded with the century, and there even seems to have been a campaign against them. An old book on flower lore warns that love-lies-bleeding draws lightning and should not be planted near a house. And a magazine of the early 1900's calls amaranths those "vulgar, gaudy plants" and dismisses them as useless, weedy, and quite unfashionable.

Amaranths Today and Tomorrow
Now, after a half century of neglect, the amaranths are staging a comeback—on two fronts.

As ornamental plants they return not under their old familiar names but as hybrid beauties called something else entirely: one nursery, for instance, lists them as "Summer Poinsettias." These are, in any event, disciplined plants that produce lush racemes of flowers and variegated foliage in a gamut of colors—salmon, scarlet, crimson, emerald, bronze, and pure white. Gardeners with a nostalgic yearning for yesterday's exotics will perhaps find them irresistible.

It is as a food crop that certain amaranths may hold exciting prospects for the future. A clue lies in a plant that grows wild in many parts of North America. Its stems are thick, its leaves coarse; its spiky flower heads, going to seed as summer progresses, turn a dark, dull brown. Small wonder that it should be considered one of our least appealing native plants and bear the common name "pigweed." But in fact it is one more immigrant, and its resounding botanical name is *Amaranthus retroflexus.*

The conquistadors brought it to the New World and grew it here, as it was grown in Spain, for a common pot herb. It escaped from early gardens and spread aggressively north, east, and west across the continent, encountering little opposition from its own kind.

The Indians soon discovered that the European amaranth was edible; the western pioneers also gathered and cooked it. And in recent years devotees of wild foods have rediscovered pigweed, whose leaves—palatable enough when young—have a higher vitamin and mineral content than either beet greens or chard. Varied amaranths have long been cultivated as pot herbs, and a variety known as "tampala" or "Chinese spinach," used for centuries in Asia, is now available from a few American seed houses.

But there are more impressive amaranths than these, and they have been grown in the New World all along. They are the so-called grain amaranths, which were a major crop in the tropical highlands of the Americas when the Spanish first arrived.

Montezuma collected a tribute of some 200,000 bushels of amaranth seed annually—almost the equal of the tribute in maize—from seventeen provinces of the Aztec Empire. And the grain had a high religious significance for the Aztecs, who carried idols made from amaranth-flour paste, mixed with honey or human blood, in ceremonial processions, at the end of which they broke and ate them. (A gentle relic of this tradition persists in modern Mexico, where a confection called "alegría," made from popped amaranth seeds bound with syrup, very much like our popcorn balls, is connected with certain festivals and saints' days.) For this reason *Amaranthus*, a symbol of heathen idolatry, was anathema to the Spanish church, whose fathers tried to suppress its cultivation and persuade the Indians to grow wheat instead. (See Wheat.)

Three species of amaranth are still cultivated—though as a minor grain crop—in the Andean regions of Argentina, Peru, and Bolivia, and in Guatemala and Mexico. *Amaranthus hypochondriacus*, the Mexican species, is the same one the Aztecs grew, and the same as the garden plant prince's-feather. But the plants bearing pale seeds are those chosen for cultivation; dark-seeded plants, like prince's-feather, are culled out because these seeds tend to produce weedy plants with a low yield. *A. hypochondriacus* has been a somewhat more widespread and important grain crop in Asia, where India's hill tribes have grown it for a century and more, and where it is now spreading to the plains. Incidentally, archeological evidence indicates that *A. hypochondriacus* was probably the amaranth that the Arizona cliff dwellers cultivated between A.D. 1350 and 1400.

Agronomists, nutritionists, geographers, and others engaged in a quest for "forgotten foods" that could become valuable new crops are now paying particular attention to the grain amaranths. Experiments inspired by Dr. John Robson, director of the Human Nutrition Program at the University of Michigan, have shown that amaranth seed is at least as rich in protein as soybeans, and contains amino acids essential to nourishment that most vegetable proteins lack. Flour milled from the seed possesses gluten, and can be mixed with wheat flour to make excellent breads. And the leaves, like those of pigweed and tampala, can be picked when young and used as greens.

Under the auspices of Rodale Research and Development of Emmaus, Pennsylvania, grain amaranths are now being grown experimentally in many sections of the country. Enthusiasts believe they may prove an ideal food crop for both the developing countries and for our own gardens and small farms. The yield per acre is relatively high, but the plants need personal supervision and the seed must be gathered by hand. If the plant wizards turn their attention to the grain amaranths, however, the resulting hybrids may increasingly help to feed a hungry world.

ASPARAGUS

There are various reasons why asparagus, a delicacy that enjoys only a brief spring season, arrived in New England with the Puritans.

An herb native to Britain and most of the Old World, it grew in every kitchen garden, from the great manor houses to the smallest cottages. It also grew wild in the fens of Lincolnshire, and in the sandy meadows of Essex, and on Cornwall's Asparagus Island. Great quantities of wild asparagus were shipped to the London markets each spring, not just because it was a delicious food, but also because it was considered an essential spring tonic, a laxative, and a diuretic. But its greatest asset for the colonists was its adaptability to many types of soil and the fact that, once established, it could be depended on to provide food without further demands on their labor.

Among the earliest New England colonists was Henry Josselyn, assistant to Captain Mason, patentee of New Hampshire. In 1634 Henry's young brother, John, joined him for a short visit. John Josselyn was deeply impressed by the American wilderness, and became keenly aware of the effect the immigrant plants were having on the virgin land.

He visited New England again almost thirty years later, armed with John Gerard's *Herball.* With this famous and venerable volume to assist his identification of plants, he spent the next eight years traveling through the region, making notes on its natural history, listing native plants as well as those brought by the English, and studying how the latter adapted to the New World and how the wilderness was affected by colonization. In 1672 he published *New-England's Rarities Discovered*, the first detailed account of American horticulture, which includes the information that the English brought "sparagrass" to Massachusetts, where it "thrives exceedingly."

It flourished not merely in the Puritans' first gardens, beside their imported mints, horse-radish, and dill. American birds quickly discovered its bright-red berries and carried the seed to the sandy meadows, marshes, and riverbanks, where it found an ideal home. An immigrant plant was off to conquer the continent.

By 1776 asparagus was growing in every colony along the Atlantic: the Dutch had brought it to New Netherland, the Pennsylvania Quakers listed it among their essential plants, and John Randolph's *Treatise on Gardening*, published in 1765, the first book on horticulture adapted to American soil and climate, notes that asparagus thrives in Virginia. (See Kale.)

It crossed the Appalachians with the pioneers and in the next century followed them over the plains. Today it grows everywhere in the United States except in areas of extreme heat. California produces 44% of our asparagus for the market and for canning, chiefly in the San Joaquin Valley, and New Jersey produces 23%. It also thrives in such diverse areas as Massachusetts, Illinois, and the Carolinas. In Idaho wild asparagus is so prolific it is considered a weed.

Origin and History

Asparagus is found in the Temperate Zones from Siberia to the Cape of Good Hope, where the asparagus fern (*Asparagus plumosus*)—a true asparagus, as its botanical name indicates, but not a true fern—and the florist's smilax (*A. aspar-agoides*) were discovered in the 1860's by Thomas Cooper, an English plant hunter. In China, where edible asparagus has been grown from the most ancient times for a medicine and food, it is dried and used the year round.

There are 150 species of asparagus, but only the so-called fern, the florist's smilax, and the edible asparagus (*A. offici-nalis*) are of economic value.

A. officinalis contains vitamins A, B, and C, and was used medicinally long before it was considered a food. The ancient Greek philosopher Theophrastus, the somewhat less ancient Greek physician Dioscorides, and the sixteenth-century Italian botanist Pietro Mattioli all recommended it for various ills. It has been used for heart trouble, dropsy, liver and kidney complaints, bee stings, poor eyesight, sciatica, and jaundice; if boiled in wine and held in the mouth it will relieve toothache. But various experiments to employ it in other kinds of projects have failed, including its use as a base for the manufacture of paper. (See Flax.)

Asparagus is a peculiar plant. It bears no true leaves, but the small scales appearing on the edible spears that push through the ground in spring are green, and perform the function of leaves. The separate ferny branches into which the spears develop if left uncut bear minute yellow flowers that ripen into the red berries whose seed is spread so efficiently by birds.

It is a plant much altered by cultivation. Wild asparagus produces twelve-inch stalks the thickness of goose quills. Cultivated asparagus, transplanted at one year into manured beds, at three years produces twenty-four-inch stalks as much as three-quarters of an inch in diameter. Old methods of forcing it for an earlier and more abundant crop exhausted commercial beds; but experimenters at Cornell and the University of Missouri have developed techniques that permit plants to recover from forcing and continuous production.

Modern research has revealed that asparagus itself is an

effective control of most root pests, such as wireworms and nematodes, which cannot endure the asparagus odor and are controlled simply by planting rows of asparagus between rows of the many vegetables preyed on by such pests. The European asparagus beetle, which arrived on root stock imported to Long Island in 1856, and the fungus-bred asparagus rust, also from Europe, which appeared in 1897, are now both under control, and asparagus is almost wholly free of pests.

How Asparagus Built a House

In central Germany, between Heidelberg and Braunschweig, is an area comprising many acres of sandy soil. Yugoslav refugees were given bits of this unproductive land after World War II, and one of them noticed that wild asparagus flourished there and was naturally bleached by the sand. White asparagus is a German delicacy, so he cut some bunches and hurried them to town, where they found a ready market. Encouraged, his family worked through the entire asparagus season, cutting the asparagus and taking it to market. In the next few years he conceived the idea of cashing in on the perennial free crop, but no one could be hired to do the backbreaking work in the burning sand. He finally was able to buy an old Volkswagen camper, and a dream of returning to his native Croatia and rebuilding his own house took shape. He loaded the camper with mortar and bricks, drove to Croatia with them, and enlisted the services of his unemployed countrymen, who gladly returned to work in Germany for the *Spargel Zeit*, the asparagus season. A quarter of a century has passed. Untold trips have been made to Croatia with bits of construction material, and dozens of migrant workers have gathered the wild asparagus in central Germany. The penniless refugee's dream house is ready for him to retire to, and a seasonal money crop has been established for his German hosts.

ASTER

Asters, which reign supreme as sophisticated hybrids in our modern gardens, are also one of our oldest cosmopolitan wildflowers. In both Greece and Rome the word for "star" was *aster*, and these star-shaped flowers, sacred to the gods, decorated their temples and wreathed their altars.

Many ancient myths tell of the aster's mystical origin. The goddess Virgo scattered stardust on the earth, and the fields were brightened by asters. The goddess Asterea, lighting the night with her star lamps, looked down on the barren earth and wept, and where her tears fell asters grew. Even the evil enchantress Medea, in a moment of contrition, scattered gay asters over the world.

Throughout Europe asters retained their beneficent character. Virgil wrote in his *Georgics* that honey was improved by boiling aster roots in wine and placing them near the hives. In France asters were called the "eye of Christ," and in England and Germany they were known as "starworts," the suffix "wort" implying medicinal qualities. For centuries aster leaves were burned to keep evil spirits at bay or drive off serpents, and a "mishmash" of asters was guaranteed to cure the bite of a mad dog.

American Asters into Michaelmas Daisies
When the first settlers came to America they found the woodland meadows colored with bright and varied asters. Indian legends, like European myths, endowed the aster with a supernatural origin and expressed love for the native wildflower that painted their fields.

The first American aster to reach Europe was taken to London by John Tradescant, Jr., in 1637. His father was one

of the underwriters of the Virginia Company and a plant
hunter of note. In his catalogue of plants published in 1656,
the senior Tradescant states that he had forty American plants
growing in his physic garden and lists the *Aster tradescantii* as
one of them. But it appears to have been lost in the influx of
foreign asters in the next decades.

From the first colonization there was intense interest and
traffic in American plants. In 1687 the German botanist Paul
Herman introduced the violet or blue New Netherland aster
from the Dutch colony on Manhattan. Carolus Linnaeus, the
eighteenth-century Swedish botanist and taxonomist who
originated the modern scientific classification of plants, translated the name into official botanical Latin as *A. novi-belgii*.
The English, in the meantime, had captured the Dutch outpost; New Amsterdam became New York, and this variety
has been commonly called the "New York aster."

In 1720 Mark Catesby took the lovely New Netherland
aster to Britain. This was important: early hybridists crossed
the New England and New York asters and created a
superior variety, which was just winning recognition as a
garden flower in 1752 when Pope Gregory revised the
world's calendar. This meant that Michaelmas Day fell eleven
days earlier, on September 29, just at the peak of the flowering season for the aster that gradually became known as the
"Michaelmas daisy."

After their flare of popularity, they seem to have been
neglected until 1870, when William Robinson, an ingenious
Irish botanist, rescued them from obscurity and produced the
first of the modern Michaelmas daisies, which remain more
popular abroad than in the United States. One hundred and
fifty varieties are said to be grown in England today, but only
a few are known here.

America has, by far, the largest number of native asters;
England can claim only one. Among the best known and most
easily identified of the American varieties are: the *A. patens*,
with its three-inch clear-blue rays and its blunt leaves that

clasp the stalk; the broad-leafed, blue-and-white *A.
macrophyllus* of the woodlands; the lance-leafed purple New
England aster of the swamplands, *A. novae angliae;* the pale-
blue, heart-shape-leafed aster of the roadside, *A. cordifolias;*
and the narrow-leafed, bright-violet-blue New York aster, *A.
novi-belgii,* which Asa Gray, America's pre-eminent botanist
and taxonomist, called, in his 1848 *Manual of Botany,* the
commonest late flowering aster of the Atlantic border.

China Asters

The China aster, which is the one we grow in our gardens, is
in fact not an aster at all, but a *Callistephus,* an annual Chinese
wildflower. All true asters are perennials.

Pierre d'Incarville, a Jesuit missionary in China, found
the *Callistephus* growing in a field near Peking in 1730. He
sent the seed to Antoine de Jussieu, a distinguished botanist
and director of the Jardin des Plantes in Paris, where it was
grown successfully. The next year seed was sent to Phillip
Miller, curator of the Chelsea Gardens in London, and to
German botanists. Its appearance suggested an aster, and so it
became known by its present name.

By 1750 it was being grown from Scotland to the Rhine,
in blue, white, reds, and purples, "to adorn Courtyards and
Parlours." By 1770 William Hanbury, cleric, gardener, and
publisher of one of the first garden magazines, reported
double asters that he called "inchanting." Horace Walpole
viewed a French garden graced with 9,000 pots of asters.
Linnaeus classified it as *Callistephus chinensis,* meaning
"beautiful Chinese crown," but it continued to be called
"China aster."

Serious hybridizing began in Germany, where the quilled
asters were developed, and so many popular varieties were
created that China asters were often spoken of as "German
asters." Germany became the center of seed production for
the next century, after which that function shifted to the
United States.

China Asters in America

China asters undoubtedly grew in private American gardens
before Bernard McMahon, a famed Philadelphia nurseryman,
introduced them commercially in 1806. In his popular *American Gardener's Calendar* he refers to them as both "Chinese"
and "German" asters.

The China aster continued to be grown as a popular
hardy annual whose essential purpose was to adorn the fall
garden. Then, about 1890, American hybridists began transforming it, and today's moderns come in a multitude of forms
and colors, with great flower heads from four to six inches
across, and varied combinations of ray and disk flowers that
nod on long-branched stems. In addition, their superior keeping qualities have made them a favorite of commercial florists.
In home gardens and bouquets they bloom from July to frost.

Botanical Bits

The asters—Chinese, European, and American—belong to
the Carduaceae group of the Compositae family, distinguished
by its tubular and ray flowers.

Asters hybridize so freely that the identification of
species depends on minute characteristics. The potential of
the ancient aster family to maintain itself in the future is assured by its productivity—there are 10,000 to 12,000 seeds to
an ounce. And there are hundreds of aster species. The
greatest concentration is on America's eastern seaboard,
where fifty-four distinct species flourish, but they grow from
Maine to Florida, and across the prairies from Texas to
Minnesota.

The majority of aster species have blue to purple ray
flowers; only a dozen or so are white.

The Lesson of the Asters

Exploring an isolated area of Tibet in 1913, Frank Kingdom-Ward—one of the last of the great plant hunters—was amazed to come upon a garden of flowers from the four corners of the world. There were sunflowers, dahlias, and nasturtiums from the Americas; geraniums from Africa; stock, poppies, hollyhocks, and pansies from Europe; and the universal aster. He learned that the local governor had bought a tin of world seeds in Calcutta and planted them as an example of the simplicity of international relations in the natural world.

BANANA

The first yellow bananas in the United States were served at an exclusive dinner party in Boston in 1875. They were offered as an exotic imported delicacy, and eaten with a knife and fork. Soon thereafter yellow bananas became a fashionable dessert.

Strange as it may seem, their history began not much more than a century ago. Cooking bananas, green and red, had been a basic food in the East before recorded history, much as potatoes are in the Western world today, but there were no yellow bananas until 1836. Out for a stroll about his plantation one morning of that year, a Jamaican named Jean François Poujot observed that the bananas on one of his trees were yellow. He tasted this unprecedented fruit, and found it edible uncooked—ripe, sweet, and delicious.

No one knows where the tree came from or why it was growing on M. Poujot's plantation: it was a sheer genetic accident. He lost no time in seeing that the tree was propagated, and was soon growing acres of sweet yellow bananas. For the discovery of what has become the most popular fruit in the world, the foundation of a billion-dollar business, Poujot was eventually awarded a prize of five dollars.

The Banana's Road to America

Bananas appear to have originated in Malaysia, but records show they were grown in India in 600 B.C., and detailed accounts in China dated A.D. 200 imply they had been grown for a thousand years in Southeast Asia. Alexander the Great made note of their use when he invaded the East some 300 years before the birth of Christ, and there is mention of them in the Koran. Eventually they spread to the eastern Mediterranean and across Africa. By 1442 they had become a basic food in Guinea, on Africa's west coast, and were ready to conquer new continents.

In that year Prince Henry the Navigator ordered his ship captains to follow the Arabs' custom of sealing fresh water in airtight casks or *toneis*—hence the term "tonnage"—which enabled the Portuguese to sail days longer, as far as the Gold Coast. There they found both bananas and the slave trade flourishing, waiting to be exploited. They entered into trade in both with the greatest enthusiasm, and the next decades saw slaves from Guinea and their basic food, bananas, flourishing in the Canary Islands, which became Spanish possessions in 1496.

Less than a decade after Columbus discovered America, a ship arrived in Santo Domingo with the first cargo of slaves, and in 1516 Padre Tomás Berlanger introduced bananas from the Canary Islands as the cheapest and most satisfactory food for the growing slave population. The banana tree adapted to the New World with the arrogant ease of a conquistador.

Francisco Oviedo, whom King Ferdinand sent to the West Indies in 1525 as inspector general, gives us some notion of the banana's success in his *Natural History of the New World:* he reports that 4,000 banana trees were cultivated on his own plantation, and that other plantations were much larger. During the next century, as we learn from many Spanish accounts, bananas made a basic contribution to Caribbean economy, and spread remarkably fast through early settlements in Central America, Mexico, and Florida.

When Florida became a state in 1819, our young Department of Agriculture showed great interest in the development of her semitropical horticulture. In 1841 it introduced Chinese dwarf cooking bananas, which had recently been brought to France from French Indochina as a delicacy for the king's table and had become immensely popular; but Americans showed little interest in them, and today only a few bananas of any sort are grown in the United States.

Cooking bananas were introduced to Hawaii by the original Polynesians, who took them there in their great outrigger canoes, along with other essential foods such as coconuts, breadfruit, and sugar. They grew and flourished in that island paradise, where yellow bananas are now cultivated for both local use and export.

The Yellow Banana

Outside Jamaica little was known of the sweet yellow banana until 1870, when a Cape Cod trader, shipping bamboo from the West Indies, included a few bunches in his cargo. He offered them for sale on the Jersey City waterfront; they had become ripe and delicious on the voyage, and found eager buyers. Word about them spread, and another trader included a few bunches in a cargo of coconuts bound for Boston. They sold quickly, too, and stimulated a demand for more.

In 1876, at the Philadelphia Centennial Exposition, yellow bananas were sold as a "Curiosity of the Indies" for ten cents each, and proved highly popular. Soon bananas became common in our ports, and the Italian fruit vendor appeared—a colorful figure on city streets, his cart hung with golden bananas, which rumor said he ripened to perfection by sleeping with the green bunches he had purchased on the waterfront.

The demand for bananas grew. The race was on.

L. D. Baker, a coastwise shipper, pioneered the first banana fleet from Jamaica. He was quickly followed by M. C. Keith, who organized a banana fleet for Central American growers. English shipping companies were soon importing yellow bananas from the Canaries to the States, and many wildcat companies and independent shippers fought to share in the growing United States trade. The last decade of the nineteenth century saw unprecedented shipping, and the yellow banana rapidly became basic to Caribbean economics.

But there were many pitfalls and failures, because there was no standard of quality and no established method of shipping. Many bananas spoiled on the way; many were picked immature or too ripe; crude handling and packing ruined whole bunches, and ailments such as leaf spot, bunchy top, and Panama disease soon spread among the Caribbean plantations.

Still, there was money in the yellow banana, and the larger shipping companies solved their problems by merging. Thus the fabulous United Fruit Company was founded, and it set in motion the rapid modernization of the banana trade, which ultimately revolutionized agriculture and economics in Latin America.

Refrigerator cars and ships were developed, international railways were built in Central America, research centers for control of disease and quality production were established, plantations were scattered over large areas from Ecuador to Jamaica in order to minimize crop failure caused by weather or blight. Ocean-going steamers plied inland waterways, re-placing native canoes that had once collected bananas from distant plantations, and eventually helicopters delivered them from the most isolated sources to banana ports in Central America.

Labor-saving methods of handling were developed. Heavy bunches are now cut into "hands" of five to ten bananas, labeled, and carried by conveyor from dockside to the holds of banana ships, in which the temperature is con-stantly controlled.

The climax arrived in 1950 with the organization of the Banana Industry Insurance Board, which established market-ing and shipping agreements among growers of the Caribbean area.

What's Behind a Name?

Linnaeus gave the banana the generic name *Musa* and called the two principal bananas known to him *Musa sapientum* and *Musa paradisiaca*. These names were not original with him, but had been used centuries before by Eastern caravans that traded with the Roman Empire. *Musa* is derived from the name of a famous Roman physician, Antonius Musa, who was personal physician to Emperor Octavius Augustus. Euphor-bus Musa, the brother of Antonius, was also a physician—to the king of Numidia, an African ally of Rome. In Numidia, Euphorbus discovered the banana, a fruit of the Far East known for its nutritional and medicinal value. He sent samples to Antonius, urging its cultivation and use in the Roman Em-pire, and Antonius promoted it with such success that it has carried the family name ever since. Octavius raised a statue to Musa in Rome, and Pliny the Elder used the name *Musa sapientum* in his *Historia Naturalis*, and declared that the sages of India subsisted on it.

The plantain or cooking banana, *M. paradisiaca*, is

widely cultivated in the tropics, and is the principal food for a large part of the human race. It was introduced early to the West Indies, but has only recently appeared in American markets, as people from the Caribbean increasingly migrate to our urban centers.

There are many varieties of banana, but the common yellow one is classified as Gros Michel and is often referred to as the "Martinique."

Banana Bounty

The banana is a monocotyledon—meaning that it bears only one seed leaf—like the palms and grasses; it is propagated as easily as the potato, and in much the same way, from "eyes" on its underground stem. It grows quickly, and may attain a height of twenty or thirty feet, bearing large, wide, glossy leaves whose overlapping bases form a kind of false trunk. As the tree approaches maturity, which may happen in no more than a year, the true stem rises from the ground and pushes up through the false trunk, to emerge at the top as a flower spike or bud. This gradually bends down, exposing from 50 to 150 true flowers, each of which may become a banana. Each bunch of bananas—a surprise to those beholding a banana tree for the first time, because the bunch grows upward from the stem end instead of down—contains between ten and twenty hands of fruit. These are ready for harvesting in about six months. The fact that they ripen best if they have been picked green is a boon to the shippers, who have learned to control ripening in packing houses and on ships with a constant temperature of 56° F. and 90% humidity.

There is another way in which the banana is a curious fruit. Seventy-five percent of it consists of moisture, and it has the ability to go on breathing after it is cut from the stalk. As it breathes it gives off ethylene gas and carbon dioxide, so the shipping bays must be ventilated to prevent the heat of the gas from ripening the bananas before they reach their market.

New Fields for Bananas

The medicinal value of the banana has long been established in the treatment of diabetes, colitis, and dietary problems, and its nutritiousness and digestibility have made it a popular baby food.

United Fruit Laboratories has demonstrated that banana skins are an exceptional fertilizer, high in potash, calcium, and chlorides—a special tonic for rose gardens. Banana fibers also offer commercial possibilities for the future, and several relatively unfamiliar varieties are gaining popularity as house plants and store decorations.

The Department of Agriculture reported in 1971 that Americans ate 4,140,000,000 pounds of yellow bananas—an average of eighteen pounds per citizen. It is the world's most nutritious fruit, as well as being the most popular because it is delicious, convenient, cheap, and abundant. From a few sporadic shipments a century ago, world consumption has expanded to 10 billion pounds of bananas a year—most of them grown in the Caribbean area.

No wonder they have changed the agriculture, economy, and politics of Latin America in the twentieth century.

BEGONIA

Begonias have no rival in America's heart. Their ornamental leaves and bright flowers pour out their varied beauty for us all, whether they are immured on winter window sills or placed around summer patios or planted in parks and public squares. What romance and adventure surround them? Where did they come from? Who brought them to American gardens?

The story begins in 1690, when France sent to her West Indies possessions a new governor, former naval officer

Michel Bégon. A passion for botany was sweeping Europe in this age of exploration and discovery, and Bégon, like many young gentlemen of the day, was an eager amateur botanist who had pursued his hobby as he sailed around the world. When he landed at Santo Domingo and saw the lush vegetation and the riot of flowers, he decided to work seriously at his avocation and become a true scientist. To his joy, he encountered a French monk, Father Charles Plumier, France's foremost botanist, who had been sent by King Louis XIV to seek out the island's botanical treasures. Soon they were deep in a project to collect and identify all the plants of the French West Indies.

Among their discoveries was a tropical plant unknown to the Old World that grew prolifically throughout the West Indies. It had leaves of remarkable colors and shapes, and produced odd-shaped, exquisite flowers in great abundance. One of these unusual plants had been shipped to Europe as early as 1650, but it had remained a nameless curiosity and was eventually lost. Now, under Bégon's leadership, many varieties were discovered, described, and shipped to France. Finally, when a new species was found in Haiti, it became obvious that this whole order was more than a novelty. It was imperative that it be identified, classified, and named. The French Academy of Science decided that the family should be named in honor of Bégon—Begoniaceae—and that the flower should be called, simply, "begonia."

The French king subsequently signalized France's accomplishments by making Bégon governor of Canada, thus ending his begonia collecting.

Begonias as a Spur to Exploration

The begonia itself was fast becoming a national favorite and an international issue. It had won instant popularity not only with botanists and florists but also among the common people everywhere. Its seemingly infinite variety, its colored foliage, its ease of propagation, and its ability to flower in garden beds in summer and on window sills in winter made the begonia irresistible to flower-loving Europeans.

England looked with envy on the fame Bégon's botanical discoveries had brought to France. Why should the French have exclusive rights to these new plants? So government officials, doctors, and missionaries in the British West Indies were admonished to do some begonia hunting. By 1717 the first begonias were arriving in England from Jamaica, and enthusiasm for them ran high. The public demanded more plants, and more. The intense interest spread across the Continent, and initiated a new period of plant exploration by European nations.

The seventeenth century had been the age of the amateur in botanical exploration. Individuals financed their own excursions, and usually their only gain was fame. But the begonia competition forced European governments to give fresh consideration to the demand for foreign flowers, ornamentals, and agricultural products; and plant exploration became a national responsibility.

By 1777 both the Spanish and the French kings had organized and financed groups of trained botanists and sent them off to Central and South America. These were the first plant explorations to be recorded as essentially a government activity. During the next century, botanists from many nations searched through the New World and the Pacific areas, not just on government orders, but also at the behest of botanical gardens and privately owned commercial nurseries. They found many new begonias; to explorers sent out by the great English nursery Veitch and Sons goes the honor of having discovered the incomparable tuberous begonias, which they came upon in Bolivia and Peru.

Begonias did not arrive in the United States until about 1880, when the Prince Nursery, of Flushing, Long Island, imported some—possibly at the suggestion of the local French

Huguenot community, which included many devoted horti-culturists. In due course the adaptability, variety, and beauty of begonias won the devotion of Americans throughout the nation.

Begonia Botany

Begonias are found in the moist forests of the equatorial zones around the world. Some grow at sea level, others in the mountains; in the Andes they may carpet whole acres. None are native to the United States or Europe, but many adapt readily to the Temperate Zones, thriving wherever they find humidity, lime, and the well-drained soil that is essential to them.

The begonia bears both male and female flowers. The male flower has four petals, the female five, and the seed is enclosed in a winged capsule. Among the numerous species and varieties, the leaves vary widely in size, shape, texture, and color. The begonia has been much hybridized, and is easy to propagate by tuber or root, and by stem and even leaf cuttings. It is also easily grown from seeds, which it bears in profusion: some of the hybrid giants are reported to produce more than 100,000 of them to an ounce.

By now some 2,000 varieties of begonia have been discovered and classified. They are divided into 350 species, besides hundreds of hybrids, but all of them fall into one of three classifications according to their root systems: the tuberous, which are magnificent outdoor summer bloomers; the rhizomatous, which as house plants tend to bloom in late winter; and the fibrous-rooted, familiar to us all, which, with careful treatment, will bloom the year around, on window sills in winter and outdoors in summer.

Among the most famous of the species grown for their ornamental foliage is the Rex, discovered in Asia and sent to England in 1858 by the East India Company. Its rough-surfaced leaves are metallic green and white, or red and purple, or marbled, or they may be a blend of all these colors. The wax begonia, or *B. sempervivens*, is the best-known class of flowering begonia; it includes the rose, Christmas, and flaming begonias.

Begonias continue to produce surprises. Some of the more recent varieties have been discovered in the Philippines, and the latest to cause an international stir is the Rieger, a hybrid developed about twenty-five years ago in Germany. It has made tremendous progress there and in the Low Countries and Scandinavia, where it is widely grown in parks, cemeteries, and public gardens. The masses of bright color created by these fabulous hybrids make an unforgettable sight.

Natives of the tropics are said to have used the begonia as a purgative and to control fever, but it serves no general economic or medicinal purposes. It doesn't need to. Few flowers can compare with the begonia—an unpredictable and enchanting floral alchemist.

BLEEDING HEART

In 1846 Robert Fortune, one of the most successful of all plant hunters, returned to London after three years in the Orient. (See Forsythia.) With him he brought the greatest shipment of living plants that had ever arrived in Europe. His triumph was due to a new device, the terrarium, which he was the first to use for the purpose of transportation of plants. Once installed in their miniature greenhouses, his specimens did not need to be watered during the five-month voyage to England. Their breath condensed on the glass walls of the tightly sealed terrariums, and dropped down again to keep them moist.

Among them was a single bleeding heart. Fortune had found this fascinating plant in the Grotto Gardens on the island of Chusan, where he had gone to see the exquisite

Weigela rosea, which grew to perfection there. Near the Weigela, among some of those artfully arranged rocks so typical of Oriental landscaping, he saw the bleeding heart, and he was able to obtain a specimen.

When it reached England it was sent with other flowers to the nursery of the Royal Horticultural Society, which had sponsored Fortune's quest in the Far East and which lost no time in ascertaining that bleeding heart adapted readily to the English climate and propagated easily. It was soon offered to the public, and its appeal was universal. In a few years it was growing in gardens large and small.

It gathered other familiar names—lyre flower, lady's-locket, Our-Lady-in-a-boat—but "bleeding heart" was the most descriptive and proved the most lasting. At the height of its popularity in Victorian days, its graceful fernlike leaves and drooping racemes of torn, bleeding hearts were used as decorative patterns in wallpaper, textiles, and embroideries.

Bleeding Heart in the United States

At the close of our Civil War, twenty years after the bleeding heart arrived in Europe from China, English nurseries began shipping it to the United States. Early American seedsmen were largely dependent on Europe's nurseries, but an unprecedented demand for plants in our rapidly growing country assured their success and forced them to become independent. The bleeding heart, so easily propagated and so appealing, soon became generally available to American gardeners, and as popular with them as it was in England.

But this was a time of pioneering. Thousands of immigrants arrived each month from Europe, and thousands of Americans on the East Coast dreamed of rich farms on the "free land" of the west. Most pioneer communities were cut off from eastern nurseries by lack of transportation, and, in any case, had no money for ornamental shrubs and flowers. Yet nowhere did men and women so long for flowers as the first settlers in this raw, new land.

To fill their need a new breed of merchants, the plant peddlers, sprang up. Often they were herb women or farmers with wanderlust, whose like has been immortalized in the tales of Johnny Appleseed. On horseback or afoot, hundreds of them wandered over the countryside, selling to eager house-wives on isolated roads and to dwellers in log cabins along western rivers. All the age-old ornamentals that could be easily grown—lilacs, lilies, daffodils, peonies—were sold. But bleeding heart reigned supreme in appeal and popularity.

One can easily imagine the pioneer woman kneeling to tend her raw prairie garden in the early spring and discover-ing a sprig of bleeding heart blooming gaily in this new land, transplanted like herself, yet enduring. She smiles impulsively, and turns to attack her day.

Bleeding heart is not so popular in modern gardens. It is thought old-fashioned. Among the bright modern hybrids, it makes little show, but it has a nostalgic appeal for those who grow it, and after five generations it is still found in seed catalogues.

America has a native bleeding heart, *Dicentra eximia*, which grows from New York to Georgia and westward to Tennessee. Linnaeus classified it from a dried and mounted specimen and gave it the botanical name *Dielytra*, the Greek word for the wing case of an insect, which the unusual petals of the flower suggested. This was later changed to *Dicentra;* the Chinese bleeding heart became *D. spectabilis*, and the American *D. eximia*, of which the two well-known varieties are Dutchman's-breeches and squirrel corn. The first living plants of *D. eximia* were sent from America to England in 1812 and were growing in English gardens thirty years before the Oriental bleeding heart was introduced. Strangely enough these American *Dicentras*, which are far less colorful and appealing, are now more grown in English gardens than the once-popular Chinese bleeding heart, which reigned so long in British borders.

A third variation of *Dicentra* is found in the mountains of our Pacific coast. But these are forest plants, difficult to cultivate, and not to be compared to the beloved bleeding heart that grew in the gardens of yesteryear.

BOUNCING BET

Bouncing Bet is a complex, fragrant, pretty flower that the early settlers brought to America as an herb for domestic and commercial use, and abandoned when they no longer needed it. Today it is the characteristic weed of railroad banks, waste places, and open fields from coast to coast.

The history of Bouncing Bet began in Europe. Its leaves provided lather for laundry and bathing, and it was brought to England from France and Germany in the Middle Ages by mendicant friars, as a gift of God to help them keep clean and healthy. Patches of Bouncing Bet sprang up around Franciscan and Dominican cloisters and hospitals, and the people of Britain soon learned to value and use what they called "soapwort."

By Elizabethan days it had become naturalized, grew abundantly along brooksides and in thickets, and was gathered by the populace for their daily needs. In an age when manufactured soap was unknown, soapwort was a blessing; its bruised or boiled leaves were used for scrubbing and sterilizing the wooden trenchers and pewter dishes commonly used before cheap china was available. It was in general use for laundry, too. It is from this that it takes its more familiar name, "Bouncing Bet"—the inflated calyx and scalloped petals of the flower suggested the rear view of a laundress, her numerous petticoats pinned up, and the wide ruff at her neck bobbing about as she scrubbed the clothes in a tub of suds.

When the Industrial Revolution brought textile mills to

England, Bouncing Bet came into its kingdom. Its roots were used to whiten and bleach the finest fabric. Its leaves were boiled to make a decoction that softened the water and removed grease, stains, and acids from the woven textiles before they were stamped with a pattern. This process of preparing or fulling the cloth gave Bouncing Bet the commercial name "fuller's herb."

As a convenience, manufacturers grew fuller's herb on the banks of their millstreams, and it soon spread from the factory districts to London's winding streets, filling them at night with a spicy fragrance that helped to mask the odors of an age when sewers were unknown. For this service appreciative citizens often called it "London pride." A perhaps less worthy function was discovered by brewers, who used it to put a good frothy head on the beer sold in English pubs.

Medicinal Use

The early herbalists all recommended Bouncing Bet, *Saponaria officinalis*, not only as an aid to cleanliness, but as a specific medicinal. John Gerard's *Herball* says it is used against filthy diseases and green wounds; John Parkinson, in *Paradisi in Sole Paradisus Terrestris*, claims that it is an admirable diuretic; and Nicholas Culpeper adds, in his *The English Physitian Enlarged*, that it is controlled by Venus and is therefore good for dropsy. Early American almanacs suggest that a poultice for boils, abscesses, gout, rheums, or jaundice can be made from two ounces of Bouncing Bet leaves and a pint of boiling water, and that a slice of the root applied to small infections has good effect.

These old remedies are substantiated by modern authority. *Kiehl's Botanical Handi-Book*—a work kept up to date by the Kiehl Pharmacy of New York—which lists the herbs, roots, flowers, spices, and berries of medicinal value growing in America today, states that Bouncing Bet is a tonic, can be used to promote perspiration, and is valuable in the treatment of syphilis, jaundice, and liver complaints.

Bouncing Bet in America

The specific date of Bouncing Bet's introduction to America is vague, but it was so commonly used in England by 1600 for dishes and laundry, and as a cure-all for skin infections, that Puritan housewives must have felt it essential for their physic herb gardens. There are, indeed, references to the settlers' dishes being scoured with "brushes of bulrushes and lather-wort," another common name for Bouncing Bet, and Geoffrey Grigson, an Englishman writing on the trials of the Puritans in *An Herbal of All Sorts*, says they suffered grievously from poison ivy, that peculiar trial of the American wilderness, but found relief by treating it with the bruised leaves of Bouncing Bet.

After the Revolution, Bouncing Bet became an important economic factor in New England's industrial development. Citizens of the new republic who had long and enviously observed Britain's thriving textile mills decided to profit from the abundant water power of their numerous streams, and manufacture their own textiles. In spite of loud protests from England, which forbade any textile worker to migrate to America and refused to export machinery for spinning or weaving, mills sprang up on hundreds of streams, and by 1840 New England had become the textile center of America. These early mills used fuller's herb much as the English factories had, and, until chemical dyes and commercial soaps were available, cultivated fields of Bouncing Bet along their streams. Today such fields often mark the sites of mills abandoned and now long gone.

Bouncing Bet took possession of even greater sections of the country when canal building began in eastern states. The canalboats provided a unique and popular way to travel inland, and inexpensive transportation for western produce back to the East Coast. Most of the canals were built and the canalboats manned by immigrants from Ireland and England, and the captains' families lived on the boats year in and year out. The women brought their soapwort from the British

Isles, scattered the seeds along the canal banks, and soon had an abundant supply for laundry and other daily use. The last of the old canalboats, moored at the headwaters of the Susquehanna and surrounded by lush fields of Bouncing Bet, have become nostalgic reminders of another day.

When railroads replaced canals as the most efficient means of westward travel, Bouncing Bet followed them across the nation and became a convenient cleansing agent for the pioneers. The steel plow furthered the process by destroying endless miles of prairie grass, which was replaced by alien wheat and immigrant weeds. Bouncing Bet, as sturdy and adaptable as the immigrant people, became naturalized along with them.

Botanical Bits

Bouncing Bet belongs to the pink family—from which, incidentally, the color takes its name, because the dominant color of these pinked or finely scalloped flowers is pale rose. Clustered at the top of stalks from one to two feet tall, Bouncing Bet's flowers are pretty, rosy, and said to be more fragrant at night than any of our other weeds. The leaves are opposite, entire, and two or three inches long. The leaves, stem, and roots all contain saponin, the element responsible for the lather that cleanses and heals.

A double-flowered form with a lavish, spreading habit of growth was grown in England in Parkinson's day and became a popular garden flower. In the Edwardian era it was used as a potted plant in homes and conservatories, largely for the sake of its spicy night fragrance. An allied variety is available today for rock gardens.

BOX

Tradition says the first boxwood in America was brought from England in 1652 by Nathaniel Sylvester, who built himself a manor house on Shelter Island and set out a box hedge around it. Thus did England put her stamp on the Atlantic coast: box was a symbol of English country life, and the use of box varied only with the climate and economy in the English colonies.

In early New England, where the Pilgrim fathers did not hold with planting for show and the climate was severe, box was little used. It could neither feed nor heal, nor did it serve any function in the house or on the farm. By 1800, the fortunes to be made in shipping and textiles were eroding Puritan disciplines, and box became the symbol of worldly wealth.

As sea captains and merchants in the China trade built mansions on the main streets of seaports, box appeared on either side of their wide front doors and grew about the sundials in their gardens, clipped and sheared into cones, obelisks, and globes. In New Netherland, Adriaen Van der Donck, lawyer, gentleman botanist, and high sheriff of the Van Rensselaer estates, wrote a book, *A Description of the New Netherlands*, published in 1656, which listed the imported flowers and fruits, and described box alleys and herringbone brick walks edged with box. In Pennsylvania, William Penn built his great house, Pennsbury Manor, on a 6,000-acre estate in the midst of the wilderness, and brought in five English landscape gardeners, who laid out a formal garden with parterres and beds of convoluted shapes, in patterns, all edged with English box, such as he had known in Europe. In our southern colonies, box reigned for two centuries, extending the fashion of the formal garden, with clipped hedges, alleys, and arbors, and with parterres to rival those at Pennsbury Manor. On the great rice, tobacco, and cotton plantations of the Tidewater, luxurious landscaped gardens ran down to docks where ships from Europe delivered exotic flowers, fruit, trees—and boxwood.

A few old boxwood gardens remain in Maryland and Virginia, including those at Mount Vernon, for which George Washington imported box, and at Gunston Hall. The charm and beauty of these formal gardens still bewitch us and boxwood's fragrance magically evokes the past.

Early History of Box

Through eons of time, box has been identified with wealth and landscaping. Isaiah exhorted the Jews to beautify the temple with box, and in Persia, Egypt, and early Greece box was essential to formal gardens. Once regarded as a native Mediterranean shrub, it is now thought to have come to the Middle East in early caravans from China. From Egypt's gardens, Greece and Rome copied topiary art—the clipping, shearing, and cutting of box, privet, and cedar to create trimmed hedges, arbors, and sculptured forms. Pliny the Elder described, early in the first century, his elaborate topiary garden, and revealed that there were master topiarists trained in the art of shearing box into amusing scenes and decorative shapes.

Topiary art, superficial and unproductive, died out in the Middle Ages, but it was reborn in the Renaissance gardens of northern Europe, where box reigned anew.

How Box Got Its Name

In ancient Greece, perfumed unguents were kept in tiny decorative boxes similar to Victorian snuffboxes. They were usually made from hard, yellow, beautifully grained box-wood, which took on a glowing sheen when polished. In Greece such boxes were called "pyxos." The Romans translated this to the Latin *buxus*, which eventually became the English box.

Since ancient times the choice compact wood of box, both roots and stems, has been used for flutes and other wood winds. The carved and polished handle of the dagger a knight wore in his belt was made from box, and early portable sundials, designed for travelers and used through the seventeenth century, were contained in miniature cubes or columns of boxwood. Until our own century only boxwood was used for rulers, and for draftsmen's scales, triangles, and T squares, but now plastics have supplanted it.

Although box was once known in the Mediterranean lands as a rare drug of the East, it never won wide acceptance for medical use, and eventually fell into disrepute. Perhaps it was too expensive—or too dangerous—and justifiably earned its reputation for emanating an aura of the grave. William Trelease, a twentieth-century botanist, notes the little known fact that box provides an acrid poison. *Kiehl's Botanical Handi-Book*, however, says that box is esteemed in the treatment of syphilis, epilepsy, and hysteria.

Mystery of Box

Since Greek gods walked the earth, box has been associated with religion and the supernatural. Greek legend says that Diana rescued a wood nymph from Apollo's clutches and changed her into the evergreen boxwood. In early Christian churches, box symbolized life everlasting, and branches of it were traditionally used in monastery and chapel on Saint Paul's and Saint Barnabas's days. In Holland, where palms were unknown, box was carried on Palm Sunday, and in English churches it replaced the Christmas greens at Eastertime. Its haunting odor became associated with religious ritual, funerals, death scenes, and graves; among the wealthy, boxwood was used for coffins and grave linings, and box bowers were built over graves to ward off malign supernatural beings.

Centuries of testimony prove that its odor affects all who experience it. To some it suggests noble aspirations; to others, fear, evil, and depression. To the Elizabethan herbalist Gerard it was "an evil and a loathsome smell."

Boxwood Botany

Common box(*Buxus sempervirens*) is an evergreen shrub or small tree. Its oval leaves average three-quarters of an inch; its flowers are numerous but inconspicuous. Apollo cast a curse upon the box tree, and said the beauty of its flowers would never appeal to man—a myth that explains why they have no sepals, no bright petals, no honey to attract fertilizing insects. Male and female flowers are borne on the same tree and are fertilized by the wind. The black seeds are enclosed in round pods, which have three projections, suggesting a tiny three-legged stool.

B. sempervirens may grow to twenty feet. The dwarf variety, *B. microphylla*, is the box used for edging, but the variety most commonly grown in America today—to whatever extent box is grown at all—is the *B. japonica*, which is bushy and dwarfish with lighter-green leaves, and which was brought to America from Japan about 1860. Variegated boxes with silver and gold leaves are also grown to some small extent. There are about twenty species of box altogether, but none thrive in the north except near the coast, or in conservatories.

Slow growing, unproductive, aristocratic, demanding expensive care, box is in fact little cultivated today. Its distinctive odor goes unobserved in our polluted air, and legends of its supernatural powers are ignored by our sophisticated, transient society. Nevertheless, it is one of our oldest ornamentals. Fossils of box have been found in Pliocene deposits, and it has always been associated with the world's most beautiful gardens: it edged the sanded paths of those in Cleopatra's Egypt, outlined parterres at Versailles and in London, bordered cool green walks on southern plantations. Who can doubt that box will continue to form a breastwork for bright flowers in formal gardens of the Space Age?

BUTTER-AND-EGGS

From June to September our roadsides and fields are bright with clumps of butter-and-eggs, whose racemes of orange-and-yellow dragon-headed flowers and sheaves of narrow pointed leaves make them one of America's most familiar weeds. But butter-and-eggs is a Eurasian plant. It was brought to America in colonial times and has served in varied ways—in both herb and flower gardens.

It is native to England, among other parts of the Old World, and there it was often called "toadflax" or "false flax," because its leaves resembled those of true flax. By Elizabethan days it had found a place in the flower garden. Gerard says it is "a most glorious and goodly flower" and recommends it for jaundice and dropsy. Culpeper tells us that Sussex farmers of the seventeenth century called it "gallwort," and added it to the water offered domestic fowls to rid them of galls.

A generation later John Rea urged gardeners to grow it in their flower beds for color, remarking that, once planted, it needed small attention. This advice seems to have been followed with excessive ardor: we soon find Geoffrey Grigson warning gardeners about "that devil of a yellow toadflax that crowds out valuable plants." He adds that in the American colonies farmers call it "dead-men's-bones."

Butter-and-eggs is said to have been introduced to America by a Welsh Quaker named Ransted, one of that large group of Welsh colonists who came to the Delaware with William Penn. He planted toadflax in his first herb garden, and when it adapted readily he shared it with other grateful settlers, who used it in a skin lotion invaluable against insect bites. This was apparently a widespread English custom; there are numerous references to toadflax lotions in old New England records. Butter-and-eggs was also used in early America to combat the swarms of flies that tortured the settlers before the invention of flypaper and screen doors. One old book tells us that ingenious farm women "constituted an excellent fly poison" by setting out saucers of milk in which toadflax had been boiled.

But the widest use of this herb, and the one that accounted for the speed with which it spread, was as a dye. In Germany butter-and-eggs had been used as a yellow dye for centuries, and the Mennonites and members of other persecuted sects from Germany who sought religious freedom in Pennsylvania rejoiced to find Ransted's herb already established there, and were soon cultivating fields of it for dyeing their homespun. As it spread through the colonies it was generally called "Ransted," a name by which it is still known in isolated areas.

The development of commercially manufactured cosmetics, insecticides, and chemical dyes negated all these primitive uses. Neglected and forgotten, butter-and-eggs escaped from cultivation, and it has become naturalized throughout North America. In many states, indeed, it has become a destructive pest, because it spreads rapidly by seeds, roots, and creeping rhizomes, invading grain fields, where it reduces production, and pasture lands, where grazing cattle are mildly poisoned by its juice.

Folklore and Legend

The history of butter-and-eggs goes back to the Middle Ages, when fables were accepted as fact. Many other folk names have been given to it, but it has always been called "wild snapdragon," after its close resemblance to the cultivated flower, with which its early history seems interchangeable. Both plants, in fact, are members of the figwort family. The cultivated snapdragon's Latin name, *Antirrhinum majus*, means "dragon's mouth." The wildflower's Latin name, *Linaria vulgaris*, means "common flax," which the leaves somewhat resemble. Like the snapdragon, butter-and-eggs is a two-lipped flower that opens its mouth when its body is pinched, revealing four toothlike stamens and a double pistil or tongue.

An adjunct to this interesting design is a spur on the lowest of the five fused petals. And here is a small mystery. Butter-and-eggs is subject to peloria, a disease that results in monstrosities, and its blossoms may develop from three to five spurs instead of one. The cause for the condition, which was described by Linnaeus, is unknown, but research shows it can be induced by one-sided illumination. Scientists hope that butter-and-eggs may provide a clue to the causes of certain human abnormalities—a possible future service by a mysterious, ancient plant.

An old Russian legend says a kind fairy revealed to an impoverished peasant that there was a rich oil in snapdragon seeds, and fields of it have been raised in Russia ever since, for the sake of its oil, which is edible and delicious.

In sixteenth-century England the seeds were widely credited with a mystic power: three of them strung on a linen thread made a charm potent enough to protect you from all evils. And a belief that any spell cast upon you can be broken if you walk three times around a plant of butter-and-eggs in full bloom endures in Scotland to this day.

Both the common name "figwort" and the Latin name of the family to which butter-and-eggs belongs, Scrofulariaceae, have a medical significance. The figworts once provided a cure for figs, or piles. And those with deep-throated flowers, including butter-and-eggs, were also thought to be a specific for throat ailments, especially scrofula, an enlargement of the lymph glands of the neck and a feared disease of the Middle Ages. Butter-and-eggs was a perfect exemplar of Nicholas Culpeper's "doctrine of signatures," a theory predating the establishment of scientific medicine, one tenet of which affirmed that a plant could cure an affliction of the part of the human body that it resembled. For centuries this wayside weed was regarded with fear and respect.

BUTTERCUP

The buttercup has been used for a thousand years to predict a future blessed with affluence. If a golden circle is reflected from a buttercup held beneath one's chin when a mystic rhyme is repeated, one is destined for wealth. Sometimes, strangely enough, there is no reflection, but this only adds credence to the ancient rite.

Such pleasant childhood necromancy is about the only use that man has found for buttercups. They are worthless as food and fodder. They have no medicinal value, no economic worth in dyes or textiles, and they are injurious to cattle and dairy products. The buttercup, as Edwin Way Teale writes, is hated by farmers everywhere because it drives out rich, soil-improving pasture grass and legume crops.

These golden flowers of our fields and roadsides are generally either the creeping buttercups, *Ranunculus repens,* or the tall buttercups, *R. acris.* They both arrived in America uninvited—stowaways among the hay in which the sparse furnishings of the colonists were packed and in the fodder for the first cattle and horses imported to this country. In 300 years they have usurped the land, giving nothing in return.

The Buttercup, Ancestor of Modern Flowers

The buttercup does have one fascinating and redeeming feature: it is a living example of the primitive flowers of prehistoric days from which the wildflowers of our woodlands, the cultivated garden flowers, and the modern hybrids have been developed.

It has a simple flower, with five sepals and petals all of which are separate instead of being fused into a single organ; the numerous and varied pistils and stamens are arranged in a spiral around a center cone, suggestive of the cones of pines and other primitive trees. It is pollinated by small bees, which can fit into its golden cups, and which, as they seek the half-exposed nectar, shake the abundant pollen from the anthers onto the pistils. This is an elementary process compared with the fertilization of sophisticated flowers such as sweet peas and orchids, in which the petals have grown together, the concealed stamen and pistils are few, and the nectar is hidden: an arrangement demanding special insects to contend with the complicated structure.

From plants like the buttercup have come, over eons of time, step by step, through mutations and forced adaptation to climates and soils, and through the changes deliberately wrought by man (that self-appointed plant wizard), the almost infinite array of modern flowers that give us such great pleasure.

The Creeping Buttercup

The creeping buttercup is a European native and the gold-flowering weed that invades our lawns and flower beds, where, true to its botanical name, *R. repens*, it rarely stands erect. Its hairy, dark-green, deeply cut leaves are thought to resemble a crow's foot: hence the common name "crowfoot," often used for the Ranunculaceae family.

The triangular seed cases, each containing a single seed, are arranged around the central cone, and as the petals drop they turn a shiny dark brown. Each seed possesses a raised edge and a curved beak—the remains of the pistil—which anchors it in the soil and so encourages germination. Its growing season is short, from June to August, but it can reproduce by runners as well as by seeds, and the nodes of the stems will root wherever they touch the soil. Inch by inch, it has followed man across our continent.

The Tall Buttercup

The tall buttercup, which is a native of Eurasia, is the one that most Americans recognize. Its long-stemmed golden flowers nod along our roadsides, in sunny meadows, and in green pastures, from May until September. It reproduces only by seeds, and therefore can be controlled if cut before it flowers. But nature has given the tall buttercup its own defenses. Although its fibrous roots cannot persist under repeated cultivation, it grows most naturally in the midst of grain and hay fields, and so cannot be cut before the grain ripens. It also thrives in pasture lands, but there it is protected by the acrid juice of its stems and leaves, which inflames the mouths and intestines of cattle, and even spoils the flavor of milk and butter if the plants are accidentally eaten by cows. Fortunately the acridity disappears when the buttercups are cut and dried with hay; but they are a poor substitute for the nourishing grass and clovers they have crowded out.

The tall buttercup is particularly adapted to compete successfully with grain crops. It can attain a height of two feet, and its large, palmate, ferny leaves, which suggest its ancient origin, are based below the center of the stem, enabling it to crowd the slender grains and enjoy the maximum amount of sunlight for itself. Its bright-yellow flowers top the erect stems, often in clusters of twelve or more. As with the creeping buttercup, each seed case around the central cone possesses a beak to attach it to the soil and aid germination.

Thus equipped, the tall buttercup has spread to every state in the union but North Dakota—an aggressive immigrant plant that has ruined countless thousands of acres of fertile land.

Botanical Relatives

It was the Roman naturalist Pliny the Elder who named the family to which the buttercups belong. *Ranunculus* is Latin for "little frog," and refers to the aquatic habitat of some of

the species, although most are adapted to a wide range of
soils.

There are more than 250 species of *Ranunculus*, among
them some of our most familiar wildflowers and cultivated
annuals and perennials: the anemones, columbines, hepaticas,
larkspurs, peonies, and clematis, one of the few climbing
species. The turban or Persian buttercup, *R. asiaticus*, which
comes in almost every color but blue, and which Louis IX of
France—Saint Louis—is said to have brought back from the
Holy Land after the Sixth Crusade, was grown in gardens
from Turkey to England, reaching the height of its popularity during the Renaissance. A giant *Ranunculus*, with
flowers three inches and more across in shades of yellow, red,
orange, and pink, is available today, and is particularly suited
for growing as a late winter or early spring flower in milder
areas of the south and the Pacific coast. The lesser celandine,
R. ficaria, came to New England with the Puritans, who grew
it as a cure for fading eyesight. Josselyn, writing about
American gardens fifty years later, reported that it grew but
slowly in Massachusetts. It has, however, long since established itself in fence corners and along roadsides as a common
weed, if not a noxious one.

A few bulbous *Ranunculus* have been considered edible.
Under stress of famine, their small tubers have been eaten as a
pot herb, after long cooking to reduce their acridity.

Perhaps especially in view of all those enchanting relatives that we welcome into our gardens, the common buttercups seem like one of nature's jokes on humankind. Millions
of golden cups offer neither food nor drink to man or beast;
their foliage may poison cattle and spoil dairy products; their
pernicious roots drain the soil of nourishment for grains and
forage crops. And yet, and yet . . . A field deep in gleaming
buttercups delights the eye, and a prophecy of gold in the
future can warm the heart.

CANDYTUFT

This familiar, white-flowered edging plant grows in the borders of millions of American gardens, but until the end of the sixteenth century it was only a little-known wildflower on the shores of the Mediterranean.

It takes its name from Candia, the ancient name of Crete, where it ran rife. It belongs to the mustard family, and like many flowers began its career as a seasoning herb. Dioscorides, a Greek doctor of the first century who traveled with the Roman army, was the first herbalist to note its use. In a work, *De Materia Medica*, that remained standard for centuries, he says that the seed of the Candia weed is used as mustard meat.

It seems to have taken candytuft about a thousand years to reach northern Europe: it is first mentioned in England by Henry Lyte, an Elizabethan gentleman who translated Rembert Dodoens's Flemish herbal in 1578, adding many notes on English garden flowers. In his *Niewe Herball* he calls candytuft "Candy Thlaspic."

John Gerard appears to have been the first actually to grow candytuft in England. He tells us that he received his seed from a Lord Zouche, who had traveled widely through the Mediterranean countries and had returned to England with many rare seeds. Of these, he gave Gerard some "Candie Mustard" that he said grew wild along the "highwaie sides in Crete or Candia; in Spain and Italie."

Gerard was enthusiastic about this versatile herb, which was apparently *Iberis amaris*, a strong-flavored variety with persistent white blooms. He declared that it excelled other seasonings, had virtue as flavoring for home remedies, and, moreover, was a comely flower for decking out the house and garden. But a half century later, when candytuft was grown in many cottage herb gardens, John Parkinson belittled it, protesting that these "Spanish Tufts" were inferior to mustard and too mild for medicine.

Candytuft's use as a cheap handy substitute for mustard, and its ability to flourish in a variety of soils, made it so popular with common folk that it was scorned by the owners of great estates and gardens. In 1679 it was rescued from its humble career. A perennial candytuft, *I. semperflorens*, was introduced to the Oxford Botanic Garden from the Near East, and proved to be unexcelled as an edging plant. Then, in 1739, evergreen candytuft, the shrubby *I. sempervirens*, was sent to London's Chelsea Gardens from Persia. These species established the candytuft's claims as a superior garden flower. The Reverend William Hanbury, a well-known eighteenth-century botanist, recommended them as ornaments for borders and beds.

The annual candytuft, *I. umbellata*, with its innumerable bright domes of flowers, and the sweet-scented *I. odorata* have become irresistible to latter-day gardeners. Many beautiful varieties, with a wide range of colors, have been developed and grown in England and America for two centuries.

Candytuft in America

Candytuft was brought to America in the late seventeenth century. The first record of its planting is at Williamsburg, which became the capital of Virginia when Jamestown was burned in 1698. The gardens of the Governor's Palace were laid out with thought and care, and floods of flowers were imported from England, including candytufts for edgings and borders.

The governor's garden was always of deep interest to Virginians, for never were more devoted gardeners gathered anywhere than in the Tidewater in the days of the great plantations: Washington, Jefferson, Custis, Randolph, and Byrd were but a few of the expert horticulturists. They kept records, devised new tools and new methods, and introduced new plants, trees, and flowers. In their gardens candytuft grew abundantly, especially the ever-green *Iberis sempervirens*, which retains to this day its popularity in winter gardens in the south.

The first seed house to advertise candytuft was McMahon of Philadelphia, in the 1806 catalogue. But long before then, candytuft seed had been sold in many colonial towns by shrewd and capable women, who found it profitable to do so.

Botanical Bits

Candytuft belongs to the mustard or Cruciferae family which takes its name from the four-petaled flowers that suggest a Maltese cross. Although yellow is the characteristic color of mustard, white is the basic color of the *Iberis*. Pink and purple varieties have long been established, and modern hybridists have developed numerous others in rich colors. Its characteristic flowers grow in trim, small heads; its leaves are narrow and diminutive. The various types grow from six to fifteen inches tall—a trait that makes candytuft ideal for bright strips of color and neat borders and suggests the folk names "candy-edge" and "candy-turf."

In small gardens in the Orient, as well as in the West, the candytuft is a beloved and dependable plant, famous for its indefatigable flowering and its adaptability.

CHICKWEED

The common chickweed in your lawn has changed little since neolithic man. Before written history it was gathered on the plains of India, as it was gathered in early Greece and in Rome, because it was an edible green that provided food through the colder months.

Nature, with her infinite ingenuity, has designed chickweed for this service by enabling it to "sleep" in cold wet weather. The stems, the base of the lower leaves, and the sepals are furred with tiny hairs. As a storm approaches, the larger, lower leaves fold up over the young top leaves and protect them, relaxing when the sun shines again to warm the tender shoots.

With such an efficient technique for living, chickweed had spread through Europe by the Middle Ages and was used not only as a pot herb and salad green, but also as a medicinal for curing rashes and other skin eruptions. In Elizabethan days it was gathered to feed falcons, because it had long been observed that flocks of wild birds sought out patches of it for winter food. It was also fed to domestic fowl, a custom from which it derives its name. Today commercial manufacturers add chickweed seed to poultry feed to stimulate the appetites of chickens raised in confinement for mass production.

Most of the early herbalists extolled chickweed as a

medicinal. Nicholas Culpeper, a London apothecary who mixed astrology with his botany, published an herbal in 1649 in which he says chickweed is under the dominion of the moon and therefore cools and reduces inflammation.

Even a modern herbalist, Florence Ransom, author of *British Herbs*, recognized chickweed as a useful healing herb. During World War II, when England was blockaded, she organized and trained committees to collect and use native plants for food and medicine. Here is her recipe for an efficacious ointment for chilblains and rashes: one-half pound of chickweed and one pound of lard boiled together and strained.

Chickweed in America

Like Bouncing Bet and dandelion, chickweed was an old and common home cure in England, and although its arrival in America is not recorded, Puritan housewives surely brought it with them to grow in their dooryard gardens. New England proved a perfect climate for the prolific alien weed, and it spread rapidly to wayside and field. John Josselyn's account of British weeds seen growing in New England after the settlers came puts chickweed near the top of the list.

Our Agricultural Research Center says that by now common chickweed has spread from coast to coast, and even through Central and South America. It grows most densely in our central eastern states, where it has spread from lawns and meadows to cultivated fields, and has become an adverse economic factor to commercial growers of alfalfa, strawberries, and nursery stock.

But modern America also recognizes chickweed as a valuable medicinal herb. *Kiehl's Botanical Handi-Book* recommends chickweed, *Stellaria media*, as an effective cure-all for ulcers, inflammations, and skin diseases. Research has found that chickweed incorporated in ointment is a useful demulcent, and country people still use chickweed poultices for rheumatism. Its healing properties are implied in its ancient name, "starwort."

Botanical Bits

Field chickweed and mouse-ear chickweed are two other Old World immigrants that have crept across our nation. They are perennials, whereas the common chickweed is a winter annual, but all reproduce by seeds and stem-rooting. All have opposite ovoid leaves, and flowers are built on the plan of five. And all belong to the pink family, a name referring not to the color, but to the pinked or scalloped edges of the flowers, which are characteristic of these plants.

Our common chickweed has five hairy sepals, longer than the five white petals, which are deeply cut. These, with ten stamens topped with balls of pollen, and the five styles, form a perfect flower one-quarter inch in diameter that suggests a tiny rayed star. The five ovaries are filled with red-brown circular seeds that ensure this chickweed's future.

Since the days of the Druids, the beneficent chickweed has fed and healed mankind. Today it thrives on millions of American lawns, its history, beauty, and service ignored and unrecognized.

CLOVER

Red clover, *Trifolium pratense*, is native to the Mediterranean and Red Sea areas. It undoubtedly fed the flocks of the Israelites when they fled Egypt and found the green pastures of the Promised Land.

From the Mediterranean, red clover wandered northward with the Roman legions, who called it "clava" because its three-petaled leaf reminded them of Hercules's club, and it spread along Roman roads through Europe like a symbol of the legions' might. It was the Romans who carried red clover to Britain, where Anglo-Saxons called it "cloeferwort" and used it for a medicinal, as the suffix wort implies.

It was later grown in the herb gardens of monasteries throughout England and continental Europe. The clover heads, dried quickly, retain their fragrance and color and make a pleasing and apparently effective curative powder. This was used through the Middle Ages as a sedative, a digestive aid, and an antispasmodic, especially in the case of whooping cough.

By the thirteenth century red clover was well established as a dependable forage crop in the Rhine valley. Albertus Magnus, a Dominican priest and a scholar with a deep interest in natural science, cultivated red clover in Cologne during the middle of the thirteenth century. In 1240 he built the first greenhouse in the history of horticulture, and gave a great dinner party to demonstrate his invention, which made it possible to grow plants throughout the winter. Unfortunately, this miniature spring in the midst of a snowbound landscape made him suspect of sorcery, but his reputation suffered no permanent damage: in 1931 he was canonized, and is now sometimes referred to as Saint Albert the Great.

By the sixteenth century red clover was cultivated over much of Italy. Europe's first botanic garden opened at the University of Padua, and Prosper Alpinus, who had been appointed to the chair of botany, took a particular interest in red clover. He carried on an extensive correspondence with English herbalists on the subject, comparing notes on his experiments with it as a medicinal, and on ways of increasing its production as a forage crop.

Gerard's *Herball* says of red clover that English cattle "do feed on the herb, as also young lambs. The flowers are acceptable to bees." The 1597 edition contained a woodcut of red clover, one of the first illustrations of it to appear in England.

A charming watercolor of red clover still in London's Victoria and Albert Museum was painted around 1570 by a Frenchman named LeMoyne, who had been befriended by Sir Walter Raleigh after the massacre of Saint Bartholomew's Day forced him to flee France.

Clover in America

By 1750 red clover had been introduced into the English colonies on the Atlantic coast and was growing on scattered farms in New England, Pennsylvania, New York, and Virginia.

From 1758 to 1761 John Adams was circuit riding in the environs of Boston. As he meandered from one court to another he meditated on the colony's economic and agricultural development, noted that good forage seemed an immediate need, and wondered how best to promote red clover to feed the growing numbers of sheep, cattle, and horses imported from Europe.

Outstanding men in all the colonies, most of whom were large landholders, felt a pressing need for a durable forage crop, but the widespread cultivation of red clover was delayed for half a century because of the Revolution.

In the end it was Lafayette who received the credit for establishing red clover in the United States. As Washington's aide he had known many prominent Americans, heard them discuss their need for forage, and sympathized, because he, too, was the owner of great estates. In 1824, when the new republic honored him for his services by inviting him to return for a triumphal visit, he brought with him a gift of many bushels of red-clover seed. Publications of the day printed glowing accounts of the incident, and this increased the demand for red clover, which proved so satisfactory that it was soon extensively grown in most eastern states.

Before 1850, farmers of New York and Pennsylvania exported clover and timothy to the Middle West and to the

Caribbean islands, but during the next decade the commercial cultivation of these crops shifted to the Ohio valley, where the growing conditions were ideal. The long hot summers, with as many as 200 days when temperatures ranged from 78° to 90° F., provided the optimum environment. Growing red clover soon developed into a minor enterprise and, for a brief period—which the outbreak of the Civil War brought to an end—Toledo, Ohio, became the world center for clover and timothy.

Conditions changed so rapidly in our growing nation that the eastern states, which had experienced a decade of unprecedented immigration after the Civil War, were forced to import 40,000 bushels of red clover annually from Europe to feed the livestock necessary to support the increasing population. This pattern continued into the twentieth century, when World War I disrupted imports and forced the United States to become independent of foreign clovers. The Midwest again became our primary producer of clover and timothy hay.

Bumblebees Are Best

With the increased demand for forage crops, our agricultural colleges began training men in apiculture, or bee-raising, for all clovers are self-sterile and must depend on bees to fertilize the flowers. The clover belongs to the legume or pea family. The oval flower head of red clover may have from forty to sixty florets, each like a miniature sweet-pea flower. Each floret contains ten stamens and one pistil, enclosed in two partially fused petals which must be opened before the stamens' pollen can be released.

For this job, the American bumblebee is best. Seeking nectar, he pushes the fused petals apart, the released pollen is showered upon his downy head, and he carries it to the next floret and fertilizes it. Because he is larger and stronger, he opens the fused petals more efficiently than the domesticated honeybees, which were brought to America by early colonists to pollinate their imported fruits and grains. (See Apple.) But bumblebees are wild and much less prolific than the honeybees. They are also diverted from the farmer's fields by the wild clovers, of which America has about eighty species, mostly variants, and none of them of commercial value. To assure maximum production of a clover crop, one or two colonies of honeybees are also assigned to each acre of clover.

Red-Clover Seed

After fertilization it takes twenty-five or thirty days to incubate red-clover seed, which is ripe when the flower turns brown. There are two seeds in each of the tiny pods that are the ripe pistils of the florets on each clover head. The mitten-shaped seeds are a twelfth of an inch long, and they may vary in color, just as red-clover flowers do, from deep purple through rose and lilac and pink to yellow or white.

The seed crop is usually the second of the year. After cutting, clover hay is cured and the seed separated, spread on the drying floor, cleaned, and packaged for sale.

Red Clover's Varied Contributions

There are 250 species of *Trifolium*, but red clover leads them all in economic importance. This short-lived annual legume is adaptable to any moist, well-drained soil. Generally, though, a late-flowering strain that avoids frosts is grown in the north, with its shorter season, and an early-flowering one, which produces both early hay and a late seed crop, is more widely grown in the south.

Few immigrant plants are as beneficial to man as red clover, which is valuable for more than hay and seed. Together with other clovers, including alfalfa, it is one of the chief sources of honey in the United States. Pale gold in color, delicate in flavor, clover honey is perhaps sold more widely than any other kind. Of greater importance is the fact that the roots of clover enrich the soil instead of depleting it, because, like the roots of all legumes, they carry nodules of bacteria that fix nitrogen from the air into compounds that are essential to the many food crops whose roots do not play host to these all-important bacteria. Clover, thus, can be considered the best of green manures. It also retards erosion.

Red Clover's Australian Adventure

Australians were deeply interested in American success with red clover. Australia is the only continent where there is no native clover, and it has little native pasture grass. The need for forage became acute as cattle, sheep, and horses were rapidly imported following initial colonization.

Red clover adapted well to Australia, but it produced no seeds because there were neither bumblebees nor domesticated honeybees in Australia to pollinate it. So American bumblebees were imported and let loose over the fields. The lack of wild native clovers meant that the bees were not diverted from the red clover; having quickly established themselves, they pollinated it more efficiently than bumblebees alone could do in the United States.

Crimson Clover: A Challenge

Crimson clover, *Trifolium incarnatum*, is also called "German" or "scarlet" clover. It was introduced from Europe later than red clover, but today is challenging it in importance. A summer annual now grown from Maine to northern Michigan, it is available in five varieties, each adapted to a different area of the country.

Once used principally as a ground cover in commercial orchards, to keep the soil moist and to provide nitrogen and green manure for the trees, crimson clover was chosen by our Agriculture Department for experiments in reseeding programs. These programs, begun as early as 1930, were accelerated when World War II cut off our foreign seed supply. By

1965 they had proved so successful that we were producing our own seed, and crimson clover stood second only to red clover in economic importance.

Filigree Clover

Another foreign clover that played an important role in our early agriculture is filigree or pin clover. Toward the end of the eighteenth century the Spanish in Mexico imported this seed direct from the Mediterranean, and planted it for a forage crop. Not long thereafter the followers of Father Junípero Serra carried it with them into California, and planted it wherever Father Serra founded a mission. In this fashion it eventually reached San Francisco, and although it was never widely grown elsewhere, it quickly established itself in the Bay Area. Today the Tiberon hills, near Berkeley, burst into clouds of pink each spring when wild filigree clover blooms—a charming reminder of an era we think of as having been far more romantic than our own.

Crown Vetch

Crown vetch is another immigrant clover that is solving an American problem. Planted along the shoulders and embankments of our highway systems, it is a highly effective guard against erosion. Pennsylvania has been using it with great success for twenty years and more. It has saved the state many thousands of dollars, survived the coldest winters, and appeared each spring to hold the soil in place more tenaciously than ever. Throughout the year the borders of the Pennsylvania Turnpike and other highways remain neat, and the fragrant pink and lavender blossoms are a particular pleasure to summertime motorists.

Shamrocks and Four-leafed Clovers

The shamrock is a trefoil, a three-leafed clover. It came to fame when Saint Patrick landed in Ireland in 433. When the

Irish ignored his Christian teachings, he picked a shamrock and used it to illustrate the unity of the Trinity and soon won many converts with this modest symbol. Since then, the shamrock has meant good fortune and is worn on Saint Patrick's Day.

The four-leafed clover—if you can find one—is guaranteed to bring you good luck, too—a superstition that has followed man from the Red Sea through Europe to the New World. It is as old as his dependence on clover, for it is a mutation of *Trifolium pratense*. In such an enduring belief there is a grain of truth. Can it be that to plant the beneficent clover is to enrich the land?

COFFEE

Surprisingly enough, coffee arrived in America with the first colonists, at a time when it remained an exotic drink in Europe, familiar only to the elite and the venturesome.

Captain John Smith, who had learned to enjoy coffee during a series of adventures that included fighting against the Turks in Transylvania and Hungary and a period of slavery in Turkey itself, is said to have brought it to Virginia. The burgomasters of what had been New Amsterdam, once resigned to English rule and the name New York, enjoyed the coffee that came on English ships as much as the tea that had come on Dutch ones.

We can sympathize, today, with William Penn, who ordered some coffee from a shipment that arrived in New York in 1683 and was outraged when he was charged eighteen shillings and eight pence—then worth about $4.65—for a pound of it. The reason for the price of Penn's coffee, and for the exorbitant price we pay now, is the same: the demand is greater than the supply. But there have been recurrent periods when the reverse was true, and it is possible that coffee may yet again become a glut on the market.

Despite its present cost, it remains the most popular beverage in the nation: surveys show that Americans drink, on the average, from two to three cups a day.

Mystery and History of Coffee

The coffee tree is native to Ethiopia, and takes its name from the Kaffa district in that country. It was early introduced into Arabia, and from the Arabian coffee tree and its progeny comes most of the world's coffee today.

Coffee was originally considered a food: for centuries Africans chewed the ripe berries, which soothed their hunger and helped them endure the forced labor that was their lot. Warriors carried rations of ground coffee berries mixed with fat and molded into a ball. This effectively supplied them with nourishment, because raw coffee beans are 12% protein, and also acted as a stimulant before battle.

When the use of coffee spread, by degrees, to Persia, Egypt, and the Holy Land, magical properties were attributed to it because of its stimulating effect, which derives from the caffeine in the beans. It became an asset to priests, dervishes, and medicine men, and increasingly gave rise to a number of myths and legends.

The brewing of coffee as a drink—a process perhaps first come upon by accident—may have begun as early as the tenth century; by the fifteenth it had become a part of the domestic and social life of the Near East. Various methods of preparation were devised, involving special equipment for roasting, brewing, and serving, and the addition of sugar and spices. It was peddled in city streets, and in Mecca there sprang up public coffeehouses, where men gathered to discuss politics, business, and the arts, or to play chess and such other games as their faith permitted them, surrounded by the fragrance of roasting beans and stimulated by the drink itself. The fame of

Mecca's establishments spread rapidly, and coffeehouses opened in Constantinople, and in Venice and Rome. Italian church fathers for a time attempted to ban them, calling coffee the drink of the infidels, but legend says that Pope Clement VII gave it his blessing as soon as he had tasted his first cup. It was he, in any event, who made coffee a Christian drink.

Coffee first arrived in England in 1650 or thereabouts, when a London merchant named Edwards, who had been traveling in the Near East, returned from Smyrna with a supply of Arabian coffee and a Greek servant who knew how to prepare it. The delighted friends he invited to coffee hours begged to bring friends of their own, until crowds besieged his mansion, demanding to try the new drink.

Edwards decided to turn his problem into pounds, and in 1652 he set up a public coffeehouse on Corn Hill, with the Greek servant in charge. It was an instant success, and the coffeehouse soon became a national institution, the forerunner of the English gentlemen's club. Every middle- and upper-class man had his favorite coffeehouse and went to it daily. There he exchanged gossip, held business conferences, or met with clients, and kept up with the world at large—for at this time there were no daily papers, so such establishments became a medium for the dissemination of news.

In time each coffeehouse acquired its special clientele. There were some where famous writers of the day could be seen and heard, and others where religion was the chief interest, for the Reformation was in full swing. And of course there were those where politics was the specialty and all the hotheads could hold forth. They became so influential that Charles II made several attempts—all unsuccessful—to suppress them.

Coffeehouses also opened in Paris, Berlin, and Vienna, and in the American colonies. Patriots are said to have "hatched the Revolution" in Boston's Green Dragon Coffee House.

The rapid increase in the consumption of coffee brought wealth to Arabia, whose port on the Red Sea, Mocha, gave the beverage a second name. Arabia forbade the export of berries or plants, but the enterprising Dutch, who controlled the East Indies, decided to introduce coffee to Java—which was to give coffee a third name. An Amsterdam burgomaster named Wieser instituted a plan for smuggling coffee seedlings out of Mocha and starting coffee plantations on Java, where the climate was perfect and labor cheap, and where the Dutch could reap a profit from both production and shipping.

Having watched the development of this enterprise for several decades, the French decided to introduce coffee to the West Indies. The only source at their command was a lone coffee tree in the Jardin des Plantes, where it received the personal care of the director, Antoine de Jussieu. One account says that grafts were taken from it in 1720 and openly shipped to Martinique. Another says that Jussieu refused to allow cuttings to be taken lest doing so destroy the tree; so it was stolen bodily and whisked off to Martinique. There, at any rate, the first coffee plantation in the New World was soon established—with such spectacular success that in a few years it became possible to supply all the West Indies with coffee trees.

Coffee plantations were well established in Jamaica by 1728, and soon Jamaica coffee was competing with Mocha and Java. In 1774 a wealthy Brazilian who was visiting a Jamaican planter became, like thousands of European gentlemen before him, bewitched by a good cup of coffee. His host presented him with a seedling when he departed for Brazil. When it throve he took cuttings from it, and later saved and planted its seeds; thus he started a plantation that was soon producing coffee for him and his friends. Other plantations quickly sprang up, not without the help of occasional smuggled cuttings, and before long Brazil was well on her way to being the world's leading producer of coffee. And all this started with one lone coffee tree!

Notes on Growing Coffee

Coffee can be easily propagated from seeds and cuttings, but it makes a number of demands. Frost can prove fatal to it, and it requires a tropical or subtropical climate with plenty of rainfall, and prefers soil rich in volcanic or organic matter. It can be grown at sea level, although the better grades generally come from groves above 1,500 feet, on up to as high as 6,000. The critical point in the art of coffee planting occurs when the young trees are transplanted from the nursery. The taproot of the seedling must be put in the ground absolutely straight and the soil firmed around it, or it will die. In the subtropics of Central and South America, seedlings are planted at the beginning of the rainy season. It is customary for coffee groves to be shaded by other, taller trees, which filter sunlight and provide shelter in general.

The coffee tree is a beautiful one, with fragrant white flowers running along its branches which blossom over a period of several months. The flowers are followed by red berries—botanically, drupes—one-half to three-fourths of an inch in size, each of which usually contains two seeds. In the East these are known as *bunns, bunn* being the Arabic word for coffee. English has corrupted this to bean—which in fact the seeds somewhat resemble.

The tree bears its first crop in three years, and after that produces three crops a year, reaching its maximum yield between its fifth and tenth years.

Growing Coffee in the United States

Since our nation is the largest consumer of coffee and the profits are so great, it is commonly asked why we don't grow our own. Coffee has in fact been grown in Florida and along the Gulf Coast. But the irregularity of its flowering, and of the ripening of the berries, compels hand picking, which results in prohibitive costs except where there is a surplus of labor. This and the fear of frosts discourage large investors, but today's high prices suggest new reflections on home-grown coffee.

A recent issue of the *Vineyard Gazette* recalls that a number of local Massachusetts farmers grew their own coffee successfully in the 1890's, and claimed it was the best-paying crop on Martha's Vineyard.

The Queen and the Commodity

Brazil is the queen of the coffee countries. She began her reign in the 1850's, when the coffee disease *Hemileia vastatrix* wiped out the rich plantations in India and Malaysia. Coffee has brought Brazil, with her ideal environment and, at one time, her supply of slave labor, untold wealth but national turmoil. When the slaves were freed in 1871 they left the plantations for the cities, and Brazil imported thousands of European colonists. When coffee surpluses caused prices and wages to drop, the colonials rebelled. Many and varied attempts to regulate coffee production, from burning surplus crops to government controls, have met with only partial success.

Now nature has taken a hand. Since 1970 Brazil has suffered a series of frosts. In 1975 a devastating one killed two-thirds of the coffee trees, and in 1976 frost attacked again. Brazil has been forced to use her reserve supply of coffee, and as prices have soared, many of Brazil's customers have turned to other sources. Central American and African countries have increased their coffee production, as has Colombia, and now rival Brazil in commodity markets.

Brazil, whose welfare and national economy have been built on one industry, may face economic disaster. What will coffee's future be?

COSMOS

Even as recently as a decade ago Oaxaca held the magic of another era. Burros bearing gay panniers laden with fruit and flowers wandered leisurely through the streets to market. Babies traveled in bright *rebozos* on their mothers' backs. Vendors in colorful serapes sold their various wares from great woven trays carried on their heads. The churches, the sunny square, the governor's palace, the absence of tourists— the atmosphere in general—made one feel transported back two centuries, to the time when Mexico was a Spanish colony.

Outside the city, Indian farms remained exactly as they must have been then: sod huts of grassy bricks, a farmyard enclosed by a living cactus fence, and along that fence tall lines of cosmos—red, pink, yellow—nodding bright heads. The last was a touch truly Mexican, which we saw many times over as we drove through the foothills of the Sierras. Mexico is a land of flowers and the cosmos a beloved native.

The Cosmos in Europe

When the Spanish conquered Mexico they found superb gardens there, and an endless variety of cultivated flowers and wildlings—a treasure that other European countries coveted. To protect it, Spain's Council of the Indies met in Madrid in 1556 to establish a policy for Spanish America, which was to affect the history of the whole New World for the next three centuries. The council recognized that the greatest treasure of Spain's vast American possessions was not the untold wealth of gold and silver and jewels, but the vegetation, from the most delicate wildflower to the tallest tree. To discourage other European nations from raiding such natural resources, the council passed a law forbidding exploration by foreigners.

It further forbade the publication of any information on the natural resources of Spanish America without a license from the Council of the Indies; in fact, such a license was never issued, even to Spain's own botanical collectors.

Then in the late eighteenth century, when interest in botany was at its height, Charles III came to the Spanish throne. Charles, who was a liberal spirit, considered his fabulous American possessions and decided to make known to the world at large the beauty and abundance of their fruits and flowers, which were still only a rumor in Europe.

He organized and financed a committee of naturalists and artists and drew up detailed plans for an extensive and intensive botanical exploration of Spanish America. In 1788 a great festival was held to celebrate the departure for Mexico, and to proclaim Spain's interest in the natural science of her New World possessions.

For the next two years a steady stream of specimens arrived at the botanical garden in Madrid. Many of them were sent on to favored aristocrats and to those who were friends of the committee. One of these was Don Antonio Cavanilles, who was famous for his garden of exotic plants. Among the seeds from Mexico planted there were those of the cosmos— the first to grow and bloom in Europe.

Don Antonio generously shared his windfall of seeds with the Marchioness of Bute, wife of the English ambassador to Spain, whose Madrid garden he greatly admired. Like the wives of many British diplomats, the marchioness took English plants or seeds with her wherever her husband was assigned, and continued her ardent gardening. In 1798, when the Butes were recalled to England, the marchioness of course carried her rare collection of Mexican seeds back to London. The cosmos was among the many New World plants thus introduced to Britain.

Lady Bute's importation of Spanish-American flowers proved to be both fortunate and fortuitous. When Charles IV succeeded his father to the Spanish throne he immediately re-established the edict against the introduction and dispersal of Spain's New World flora. He also canceled support of the plant hunters in Mexico, leaving them stranded and penniless, and their discoveries ignored.

The cosmos, however, adapted well to the English climate, flowering prolifically and providing seed for other great gardens. But it was not cultivated for commercial sale, or made available to the average garden, until the middle of the nineteenth century.

Cosmos in the United States

This is probably why cosmos was unknown to American gardens for so long. European nurseries had customarily shipped new flowers to the States as soon as they had a surplus. But the confusion of trade caused by our Civil War, and America's absorption in pioneering, diminished the importance of European flower seeds and nursery ornamentals and opened the way for the growth of American seed houses. None of them listed cosmos. Indeed, cosmos is not even mentioned by Asa Gray in the early editions of his *Manual of the Botany of the Northern United States*.

Nevertheless, the cosmos was to have its day in American gardens. A new surge of flower breeding was on its way, and it culminated in 1897 with the founding of our Plant Introduction Center in Washington, D.C., which sent agriculturists all over the world to search for plants and flowers of economic value that would be adaptable to our varied climates. An initial project of the center was a quest for plants adapted to the arid climate of the southwest, and Mexico was a nearby, natural place to look now that the Mexican revolution had ended the restriction on exports.

The first cosmos was introduced directly from Mexico in 1898. It adapted quickly under cultivation, achieved almost instant fame, and has since become one of the most popular of all the annuals that came to us from Mexico.

Botanical Bits

The cosmos belongs to the aster branch of the Compositae family. There are twenty species known, and all are native to tropical America. It has borne its Greek-derived name since the Spanish padres first grew it in mission gardens, noted its ordered simplicity, and saw in it a symbol of harmony; so they used for it the word that means an ordered universe. Its aptness and its own simplicity seem to have obviated the need for any common name.

The two species grown in American gardens are *Cosmos bipannatus*, with petals of rose, purple, and white surrounding yellow disk flowers, and the *C. sulphureus*, with sulphur-yellow petals and a bushy center disk.

Modern plant breeders have transformed the Mexican wildflower, and new hybrid cosmos makes a magnificent background for today's gardens, with plants ranging in height from four to six or more feet, bearing in a season as many as a hundred bright flowers two or three inches across. Their somewhat fragile stems and delicate, threadlike leaves enhance a look of great airiness and elegance.

Although the cosmos reveals its origin by a love of hot sun and an ability to thrive on neglect, it is a superb flower for both planting and cutting. It transplants well, and can be grown as a tender annual in the north; in the south it blooms without cessation from early spring until frost.

It has an added—and rare—asset. It will bring humming-birds to your garden.

CRAB GRASS

Please don't kick the crab grass. It's good for man and beast, and that's why it's growing so persistently on your lawn. You may not like it there, but thousands of years before lawns were dreamed of, crab grass was grown for food. It was, indeed, the first grain cultivated by man.

The Stone Age lake dwellers in Switzerland grew it, and the form known as "foxtail millet" was grown in China in 2700 B.C. and regarded as an important source of food. It has been grown from prehistoric times in India and Africa and used there for making a tasty porridge and bread. Even today, in some parts of the world, some types of crab grass are considered a leading cereal, and their seeds are still gathered and cooked as a staple.

There are 500 species of crab grass known to Africa and Eurasia, but not one was growing in America when the white man came.

An Official Welcome for Crab Grass
Crab grass was first brought into this country in 1849, by the United States Patent Office, which at that time did the work since taken over by the Department of Agriculture.

When the first settlers arrived there were no cattle, sheep, hogs, or horses in America, and much of the east was covered with virgin forest. The rapid growth of our country, and the importation of thousands of domestic animals, resulted in an increasingly acute need for good forage. To fill it, the Patent Office, after due consideration, chose to introduce crab grass to American farms and pastures. But there was no agricultural organization to follow up the introduction with demonstrations and promotional information, and after a few ineffectual news items and local lectures, crab grass was forgotten.

So, after all, American lawns were safe for another generation.

Pioneering with Crab Grass
The real invasion of crab grass came with the wave of immigrants from Poland and central Europe, which began in the

late nineteenth century and reached its peak in 1910. With the opening of free land in the west and the rapid industrialization of the east, a great influx of Poles, Czechs, Slovaks, and Hungarians poured into America. Ignorant of what they might find in this new world, most of them brought along seeds they knew would grow anywhere and produce food for their families. To many of them the essential grain was a variety of crab grass. The name they called it—"manna grits"—suggests the store they set by it.

Crab grass is a form of millet native to central Europe. A feature that particularly recommended it to the immigrants was that it could be planted late in the season and still produce an excellent crop. The pioneer knew that even if he was slow in finding a claim he was still sure of a crop, because crab-grass millet thrives in many types of soil, grows rapidly, and invariably produces a rich harvest of grain on its long seed spikes.

Crab grass grew, thrived, and produced bountiful bins of grain in the new land. But soon these immigrants saw all about them great fields of wheat and corn, which were raised and harvested with far less work and which sold for far more money. No one wanted their surplus millet seed, so they abandoned crab grass, and within a decade they, too, were harvesting wheat and corn.

But the damage was done. The very features for which the immigrants valued crab grass made it ultimately hateful. The seed escaped to roadsides, wastelands, and abandoned claims. Today it has spread to every state in the nation and is the number-one pest on countless lawns.

But the story has, if not a happy ending, at least a silver lining. Our southern states needed a good forage grass for their expanding cattle business, and found that the large crab grass, which thrives in hot, dry conditions, was ideal. Under cultivation it grows to a foot or two, like other grains, and it makes a highly nutritious hay as well as green pasturage. Today it ranks second only to Bermuda grass, which it resembles, and as a hay crop it is of substantial value.

Botanical Bits

The two principal crab grasses found in the United States are the large crab grass, *Digitaria sanguinalis*, and the smooth crab grass, *Digitaria ischaemum*. *Digitaria* refers to the fingers or seed spikes, from which it takes the common name "finger grass"; *sanguinalis*, to their blood-red color; and *ischaemum*, to the arrangement of seeds along the spike's midriff. Crab grass is sometimes called "panic grass," though not because of its effect on lawn-proud householders. It belongs to the genus *Panicum*, which takes its name from the Latin *panicula*, diminutive of *panus* or tuft.

There are 500 species of millet. They vary, but most have from three to twelve fingerlike spikes, which grow in a whorl at the top of the stem. They produce minute flowers, and later bear hundreds of minute seeds. One pound of crabgrass millet could contain as many as 100,000 seeds.

No wonder the immigrants gave it up for cornflakes!

CROCUS

The spring crocus, *Crocus vernus*, arrived in England from the Middle East in the sixteenth century and caused a sensation at the Elizabethan court. Shakespeare knew it, Sir Francis Bacon grew it, and the herbalist Gerard described it. It was soon widely known, and was commonly called "Valentine's flower," because it was often in bloom by that saint's day.

The *Crocus vernus* had been introduced through Charles de Lécluse, a Flemish botanist also known as Carolus Clusius, whom Maximilian II, the Holy Roman emperor from 1564 to 1576, had appointed to supervise the Imperial Botanical

Garden at Vienna. During his fourteen years there, he sent the spring crocus to botanical gardens in Nuremberg and Paris, and also to Leyden, thereby helping to found Holland's bulb industry. It was a French botanist named Robin, as Gerard relates in his *Herball,* who sent the new crocus to him. He calls it the "spring saffron" and describes it as "one sort hereof with purple or violet colored flowers."

The saffron crocus, *C. sativus,* had been cultivated commercially in England for centuries, and the meadow crocus, *Colchicum autumnale,* or colchicum, was a native. Both these crocuses were common in English gardens, and they as well as *C. vernus* were ready to conquer new worlds when England began to colonize the Atlantic coast.

There were no crocuses in America at the time Jamestown was settled. There were no gardeners among the adventurers, either. They were all men, and they quarreled, starved, and exhausted whatever surplus energy they had in a search for gold. Two hundred thousand pounds were spent to found the colony, and laws were passed to force the colonists to plant food. But all was in vain until the women came, when everything changed. It is said they brought papers of seeds and the crocus's small bulbs packed in the pockets of their skirts and aprons. And doubtless they did, because Englishwomen have always taken their gardens to Britain's far-flung colonies. No other nation but Rome carried its gardens with it, or held so much of the world for so long.

In time other colonies were established, and crocuses grew on tidewater plantations in Maryland and in New England's harbor towns. But they were still rare and coveted in 1740, when Peter Collinson, an English merchant and botanist, shipped an order to the famed Quaker gardener John Bartram in Philadelphia that listed: "Crocus, Narcissus, Iris, Gladiolus, Martigan Lilies."

William Bartram carried on his father's garden, and his diary contains a postscript to these imports: "January 15,

1802—*Crocus vernus,* snow drops, narcissus, tulips above ground."

The crocus is a complex genus, with probably seventy-five known species. The early botanists divided them into merely two: *C. vernus,* the violet spring crocus, mother of the blue and white Dutch crocus, and *C. sativas,* the saffron crocus of commerce. Gerard describes the yellow *C. aureus,* parent of the yellow Dutch crocus, the *C. lutens.*

The Saffron Crocus

The saffron crocus, *C. sativas,* is a fabulous plant, with so long and varied a career of service that its origins are lost. It is thought that the Mongols carried it to China, and the earliest caravans to Asia Minor. In Greek myth Mercury creates it from the blood of Crocus, Europa's son, whom he has accidentally killed. More mundane proof of its presence in ancient Greece is the cyclidic jar unearthed in some ruins in Crete and judged to have been used in 1500 B.C., which was plainly decorated with crocuses.

The Greek name for it, *krokos,* or saffron, refers to the aromatic orange stigmas of the flower, which when dried and ground provided a yellow powder used to spice and color food. In powders and pills it was also used as a nerve tonic and an opiate, and for hysteria and melancholy.

The cultivation of saffron was well established in Rome before the birth of Christ, and Pliny the Elder notes that it was an important product of Sicily by the year A.D. 1. It became a valuable dye in Europe, as it was in the Orient, producing golden-yellow cloth, but only for the exclusive use of the rich and noble, inasmuch as it took 4,000 flowers to produce an ounce of saffron. This tremendous commercial importance encouraged many adulterants of saffron as both a spice and a dye. The most commonly used for centuries was the golden calendula. (See Marigold.)

The Romans had Latinized *krokos* to *crocus,* and Roman legions spread the saffron crocus throughout their empire, including England. Though it disappeared when Rome withdrew, it was firmly established again by the Crusaders, who brought it back from the Holy Land. By the twelfth century it was happily growing in the damp soil of southern England, and by the 1300's Saffron Walden had become the center of the dye industry, which flourished there for 500 years. Saffron also maintained its fame as a spice for food and drink through the Middle Ages, and the people clung to their faith in its therapeutic value.

After the development of chemical dyes in the last century, the use of saffron was drastically limited. And although it is still popular in southern Europe and the East as a condiment, it runs a poor second to artificial substitutes.

The saffron crocus was never grown commercially in America, but found a place in the early physic gardens and many flower gardens. Its fall from fame is suggested in Leonard Meager's *The English Gardener*, a horticulture handbook used throughout the colonies in the seventeenth century—Governor John Winthrop possessed a much-thumbed copy—which simply lists both the saffron and the colchicums under "Divers other pretty flowers to furnish the garden."

The Colchicums, C. autumnale

The mauve colchicum, which is called the "autumn crocus" and "meadow crocus," is not a crocus at all, but belongs to the lily family, whereas the true crocus is an iris and native to England and Europe. Colchicum has no commercial value, but, rather, a suspect reputation, being poisonous to man and beast. Theophrastus, who wrote the first known European botanical study, *Inquiry into Plants*, tells us that in ancient Greece lazy slaves ate colchicums to sicken themselves and so avoid work.

In England varied strains were grown by Gerard, Rea, and Parkinson. They still grow wild in profusion in Herefordshire, and have long been a beloved garden flower called "naked ladies" because they bloom without their leaves, which appear only in the spring.

Suspect Character and Modern Use

The roots of the colchicum are a source of a narcotic and poison, but were used medicinally for gout from the days of the Pharaohs to the time of King James I. Modern research reveals that colchine derivatives can change sterile hybrids to fertile ones, and can induce duplication of chromosomes, resulting in new races of plants.

Usurping the name "autumn crocus," the colchicum has emanated an aura of mystery throughout the centuries. Its botanical name was derived from Colchis, a city in ancient Greece where the flowers grew profusely, and where Medea restored youth to the favored and poisoned her enemies with its potent roots.

The history of the versatile crocus has a kind of parallel in the life of its historian, a tile manufacturer named George Maw (1839–1912), who became so interested in clay, and in geology in general, that he was chosen for a botanical expedition to the Atlas Mountains. This led him to horticulture, and to collecting crocuses in Europe and Asia. He was, as it turned out, an able artist. His superb drawings of crocuses are preserved at Kew Gardens and are described as a landmark, and his *Monograph of the Genus Crocus* is one of the most conclusive works ever published on any single genus.

CUCUMBER

Along the garden path, the cucumber peers out from its green cave of leaves, its emerald epidermis studded with diamonds of morning dew. Through what arduous adventures has this mysterious foreigner traveled to my New England garden?

Cucumbers are native to India and Egypt. Some forms

are still found wild in the foothills of the Himalayas, and in Egypt they were cultivated for food and drink before written history. The Pharaohs who built the Pyramids fed their slaves leeks, onions, and garlic to prevent sunstroke, and cucumbers to quench their thirst. When the Israelites escaped from Egypt and sojourned in the wilderness, they longed for the coolness of the cucumbers that grew in Egyptian fields.

Centuries before the days of thermos jugs, travelers in desert caravans carried cucumbers with them. The green skins efficiently protected the cool fresh liquid within, which could assuage thirst; the flesh provided a refreshing food. They spread through the southern Mediterranean littoral, and to China and other parts of the Far East, before the Christian era, largely because they were a necessity to caravans. And both their common and botanical names suggest this use. "Cucumber," *Cucumis sativus*, as well as Cucurbitaceae, the name of the family to which it belongs, is derived from the Latin word for "gourd"—and certain gourds have been made into drinking cups since the most ancient of days. All gourds, squashes, and melons are members of the cucurbit family.

The cucumber was introduced into Europe by Alexander the Great. Centuries later, Julius Caesar exalted it in Rome after his Eastern campaigns. Both men were naturalists at heart and saw to it that a botanist was in the forefront of their armies, to collect the plant treasures of foreign countries. In Greek and Roman literature the cucumber is praised as an essential for garden and guest, a sustaining food for the serf, a salad for the patrician's table.

By the reign of Emperor Tiberius, cucumbers had won medicinal recognition. When he fell ill, his doctor ordered him to eat a fresh cucumber each day. His gardener, confronting the winter months ahead, devised the first known cold frame, a shallow pit covered with sheets of mica. It was a great success, and mica was used for cold frames until the eighteenth century, when sheet glass became available.

Roman soldiers carried cucumbers to England. They

flourished there until the fifteenth century, but were lost to cultivation during the long, devastating Wars of the Roses, when gardens were abandoned and orchards destroyed. But Catherine of Aragon, the Spanish bride of Henry VIII, so longed for the green gardens of her homeland that the king ordered cucumbers and salad greens from Flanders, and Flemish gardeners to grow them again in Britain.

Cucumbers in America

Queen Isabella insisted that cucumbers be among the first vegetables planted in the New World, and by 1494, as Columbus reported, they were already thriving in what is now Haiti.

The Indians, who were quick to discover the excellence of cucumbers as both hunger and thirst quencher, spread them rapidly throughout the Caribbean islands, an area they clearly found congenial. It did not take them long to reach the mainland, after which they may have made their way north with the utmost speed. Fernando de Soto was amazed to find them growing in Florida when he invaded it in 1539. After having eaten some, he pronounced them better than those grown in Spain. But four years earlier—or so the story goes—when Jacques Cartier explored the Gulf of Saint Lawrence and went on to discover the Saint Lawrence River, he saw cucumbers growing in the Indian villages along its banks.

We know from the copious notes that Captain John Smith wrote about the Virginia venture that the English planted cucumbers at Jamestown in 1609. John Rolfe, shipwrecked off Bermuda on his way to Virginia, planted them there in 1610, and saw them flourish beyond all other English vegetables with the possible exception of potatoes (which see). In 1635, in *A Relation of Maryland*, Lord Baltimore's brother-in-law William Pearsley reported that cucumbers and watermelons were thriving at Saint Marys. And William

Penn's Quakers planted cucumbers in their first Philadelphia gardens, and maintained their reputation as superior horticulturists by developing new varieties for pickles and salads.

Isabella Beeton and the Risky Cucumber

Mrs. Isabella Beeton, whose formidable *Book of Household Management* first appeared in 1861 and swept Victorian England, displays a certain ambivalence about the cucumber.

"It is a cold food, and difficult of digestion when eaten raw," she remarks, and adds, to our latter-day astonishment: "As a preserved sweetmeat, however, it is esteemed one of the most agreeable." Again, she observes, "Generally speaking, delicate stomachs should avoid this plant, for it is cold and indigestible."

But besides supplying, as her custom was, a good deal of historical and botanical information, she offers a variety of cucumber recipes, though not one for that most agreeable preserved sweetmeat. Cucumbers stewed in a good brown gravy might not appeal to contemporary palates, with or without the suggested addition of a lump of sugar to take away bitterness, but there is a recipe for French cucumber soup of which Julia Child might approve. And of sliced cucumbers in an oil-and-vinegar dressing, Mrs. Beeton remarks that they are "a favourite accompaniment to boiled salmon . . . a nice addition to all descriptions of salads, and . . . a pretty garnish to lobster salad."

Eat cucumbers, then, at your own risk.

Plant Wizards and the Co-operative Cucumber

Modern hybridists have been unable to resist the possibilities offered by the adaptable and productive cucumber. They have examined its sex life, altered its chromosome pattern, and redesigned it in a number of ways to fit today's economic picture. Here are some of their discoveries and what they have led to:

✑ The intensity of light can produce great differences in cucumbers. Under bright light the proportion of male flowers is high and the crop small; but shade increases the number of female flowers and results in a large crop.

✑ Vines can be reduced to a minimum and still produce the same number and size of cucumbers per plant. Such vines take up far less room, and are as welcome to home gardeners with limited space as to commercial growers.

✑ Vines have been developed to produce only female flowers, and provide an extraordinarily heavy yield.

✑ And variety! There are pickling cucumbers that mature in fifty days from planting. There is a special hybrid slicer for greenhouses, whose blossoms need not be pollinated by bees, with a skin so delicate it need not be peeled, and it sells for premium prices. There is a Japanese strain that produces fruit fifteen inches long, or longer. And there is a strain called burpless—not a word Mrs. Beeton would have used, even if it had existed in her day, but such cucumbers might have won her wholehearted approval.

By the 1970's all this added up to a $99,556,000 crop in the United States, where 143,000 acres were devoted to cucumbers. Cultivation is concentrated around the Great Lakes, and in the Gulf and Pacific coastal regions, where thousands of acres produce 100 bushels each year. And miles of greenhouses in various parts of the nation force large crops for the winter marketplace.

The genius of the plant wizards and the amazing adaptability and fertility of the cucumber are heartening portents of a green world with food for all.

DAHLIA

As Cortez marched his army from Veracruz to Tenochtitlán in 1519, he was amazed at the beauty of the Indian cities through which he passed. The roadways and streets were lined with hedges of a strange native plant, growing four to six feet tall, that bore an abundance of dark-green leaves and small blood-red flowers. His restless eye saw, admired, and wondered, and he classified the flower as one to collect and send to Spain.

So it came about that the first exiled tubers of a plant the Aztecs called "cocoxochitl" were planted in a Spanish monastery garden. They thrived under devoted care, put up their tall thick stems and intricate leaves, and their small, daisylike, dull-red flowers. Surrounded by the extravagant beauty of the various tropical exotics that had arrived with it, however, the cocoxochitl aroused little interest, and in a short time was neglected and forgotten.

In 1571 Philip II of Spain sent his personal physician to Mexico to collect medicinal and ornamental plants. The good doctor spent six years there, writing and illustrating the first thesaurus of New World plants. In it he included a drawing and a description of the cocoxochitl, and reported that in Mexico City, besides the single flowering red species, there were several hybrids, with large double flowers in brilliant colors, that would surely prove welcome in Spain. But the doctor's thesaurus remained in the royal archives for nearly a century before it was published, and played no part in making the cocoxochitl generally known.

Diplomatic Dirty Work and the Dahlia

Meanwhile, France was badly shaken by another of Spain's Mexican triumphs.

There is an exquisite crimson dye called "cochineal" that happens to be made from the dried bodies of the female of a species of wood louse. It was known in ancient Egypt and Arabia; down the centuries it had become—because of the long and complex process of producing it—one of the cost-liest imports to Europe from the Middle East. Now Spain discovered an abundance of cochineal lice in Mexico.

The French found it maddening that Spain, which had come upon such stores of gold and silver and such a wealth of plants in America, should also be blessed with a Mexican cochineal louse. So their minister to Mexico was instructed to steal a few hundred live specimens and ship them home, to serve as the basis for a French cochineal industry.

Now the minister, more politician than naturalist, knew that the lice thrived on plants, and that they must have food for their long transatlantic voyage, but he did not know much about Mexican verdure. The cocoxochitl, however, was familiar to him because of its use in public places, so he packed the stolen lice in cocoxochitl, complete with roots, and sent them off clandestinely to France.

As it happens cochineal lice feed on only one plant, the opuntia cactus, and by the time the stolen lice reached Paris they had all starved to death. The cocoxochitl plants were alive, however, and were sent to the Jardin des Plantes. These first specimens of a strange plant never before seen in France multiplied rapidly under cultivation, and drew so much attention that surplus tubers were offered for sale to French gardeners. They quickly became popular, and ten years later were exported to England, where they received an equally warm welcome.

Holland got into the act by accident, too. In 1782 a Dutch importer who was determined to profit by the plant's growing popularity arranged with a friend in Mexico to send him a camouflaged box of cocoxochitl tubers. The friend, whose enthusiasm outran his knowledge of plants, packed them so carelessly that by the time they reached the importer they were all dead but one. But the lone tuber was planted, and when it bloomed it proved to be a new strain of cocoxochitl, hitherto unknown in Europe, bearing a large, blood-red flower with long, pointed petals that recurved toward the center. It was immediately crossed with other hybrids, and the modern plant that we now call "dahlia" was born.

A New Name for Cocoxochitl

Spain, ever jealous for her American possessions, became aware of the rising interest throughout Europe in the cocoxochitl, which other nations possessed illegally. In 1789 the king held a celebration to confirm Spain's discovery of the plant and official right to it. What better way than to change its name? As a Spanish salute to contemporary science, he decreed that it was to be called "dahlia," in honor of Anders

Dahl (1751–1789), a Swedish botanist who had developed a number of cocoxochitl hybrids. But all this only intensified French and English interest in it, and before long the dahlia was recognized all over Europe as perfect for planting in parks and other public places—just as the Aztecs had done.

By 1834 dahlia hybridizing was well under way. The demand for the showy flowers, with their huge heads, long stems, and brilliant colors, soon outran production, and dahliamania spread across the Continent; prices skyrocketed, public and private trading increased, and fortunes in tubers were made and lost. (See Tulip.) When at last the market was saturated, a new outlet was needed, and late in the nineteenth century nurserymen began to ship European descendants of a Mexican wayside flower to the United States.

American Dahlias

The European monopoly on dahlias was broken by Peter Henderson, a Scotsman, who had opened a seed store in New York City, before the Civil War, originally as an outlet for European nurseries. But the war cut off his transatlantic supplies, and he began to grow his own seeds, and then his own bulbs, and finally his own dahlias.

When, with the opening of free land in the west, the United States became a great agricultural nation, Henderson responded by establishing a mail-order business. Many other seed houses sprang up and became famous during this period, and many of them imported dahlias from Europe. But as dahlias took the nation by storm, other American nurseries began to grow and hybridize them; there are now some 3,000 varieties. The five most popular types are the single dahlia, the peony dahlia, the cactus, the pompon, and the large decorative double dahlia, with its infinite numbers of ray flowers. Although the showy dahlia reigns as queen of the late summer garden, the small-flowered dahlia with maximum bloom is used for public planting in Canada and the United States, just as it is in Europe and Mexico. This is particularly true in our south. Today's high labor costs have curtailed public planting of dahlias in the north, where the tubers must be dug and stored in the fall, and replanted in the spring. Beauty yields to economy.

It is the tubers that have helped the dahlia to endure through many civilizations. Large and fleshy, they are storage roots rather than true roots, and will not grow unless an unbroken stem complete with a bud is attached. They contain fructose, a sweetening agent that is harmless to diabetics; they are wholesome and highly nutritious, and have been used as food for centuries in Mexico, Central America, and parts of South America. It is a curious additional fact that because the Indians had grown and hybridized the cocoxochitl long before the arrival of Cortez, no one is sure what the true wild plant was like.

Born in Mexico, developed and exploited in Europe, a late immigrant to the United States, the dahlia has re-entered America in the fullness of its glory, and bestows spectacular beauty and variety upon our parks and gardens.

DAISY AND CHRYSANTHEMUM

There were no daisies in America when the Puritans unpacked their sparse furnishings in Massachusetts, but when they threw out their packing straw they unknowingly planted this gay and best-loved wildflower in our summer fields. Tradition says that the oxeye daisy crept out of Governor Endicott's garden between Tuesdays, and claimed the whole nation in two centuries. A classic example of littering!

Those who visit Brandon, one of the oldest Virginia

plantations, and see the acres of oxeye daisies across the James River, are reminded that homesick women in New England and the other colonies cherished this weed from English fields in their first gardens when they found that their English daisy, *Bellis perennis*, did not flourish in America.

The speed with which the oxeye daisy has raced over our own country explains its successful journey from China, its native land, across Asia, North Africa, and Europe. By the end of the Middle Ages it was well established in England, and ready to conquer a new continent.

Most Americans believe that this migrant daisy, "scattered as gaily and thickly as confetti," is a native flower. One modern naturalist attempted to correct the error by claiming that the first oxeye-daisy seed was brought to America in sacks of grain shipped from England to feed General Burgoyne's horses during the campaign that culminated in his defeat at Saratoga. Perhaps there is at least an ironic truth in this, because the daisy did take over the Hudson valley after the Revolution, but a disgruntled Burgoyne went home to England instead, served for a time in Ireland, presided over the impeachment of Warren Hastings, and became a successful playwright.

The Daisy—Universally Beloved

Our field daisy, whose history goes back to Far Eastern legends, has been beloved by prince and peasant in the Orient and the Western world. Its endless folk names in many nations attest to the affection that has been bestowed upon a weed that has given man no "meate or medicine," and no fodder for his cattle.

In the Rhineland it was called "Johanneskrut," the flower of Saint John's Day, a name with which the early church overlaid the pagan magic connected with Midsummer Eve, when the daisy was in full bloom. Most Germanic tribes had credited it, because of its golden eye, with the power to ward off lightning during summer thunderstorms. (See St.-John's-Wort.)

In France, Margaret of Anjou loved the oxeye daisy so well she had it embroidered on her own robes and on the raiment of her courtiers and attendants before she sailed to England in 1445 to marry Henry VI. He proved to be a feeble monarch, but this valiant, strong-willed, and autocratic queen led his Lancastrian troops in the Wars of the Roses, and the daisy flew on her banners. Thus it gained its name "marguerite"—which, by an odd twist, has become the familiar name now usually given to the garden daisy, *Chrysanthemum frutescens*, and to the cultivated forms of the true English daisy, *Bellis perennis*.

In Scotland the oxeye became the "moon daisy" or "thunder daisy," and even today people in rural areas believe, like the old Germanic tribes, that it has benign powers against thunderstorms.

Botanically Speaking

The oxeye, a member of the Compositae family, is designed with marvelous symmetry. It consists of a golden disk in which several hundred minute tubular flowers are pressed together on a cushion base and arranged in nature's dominant spiral plan, so that each will receive maximum sunlight. Around this yellow crown is a circle of twenty to thirty white ray flowers, each with a stamen and pistil. Linnaeus aptly named it *Chrysanthemum leucanthemum*, which means gold-flower/white-flower.

The Crab Spider's Home

Nature, with consummate efficiency, assigns the oxeye daisy a unique service. It is the home of the crab spider, which—small, almost translucent, colored yellow and white—is perfectly camouflaged in the center of the blossom. Here she spins her web and waits for any insect who may come seeking honey. It is a perfect booby trap, providing her sustenance with maximum safety and minimum work.

When the oxeye withers in midsummer, the crab spider instinctively moves on to a new home on the flower of the black-eyed Susan, which blooms into the fall, and there this chameleon spider takes on the flower's deep yellow and brown and continues to trap her prey until frost.

The oxeye daisy has bewitched man by its very exuberance. Although it is classified as a pernicious weed and ruins thousands of acres of productive land each year and its *raison d'être* seems to be merely to provide delight with summer fields of gold and ivory, we view it with a peculiar tolerance and have permitted it to encircle the world. It has even been brought into some modern gardens, and under cultivation has produced larger and more beautiful flowers. But, like most of the 160 species of chrysanthemum, it has remained a weed, with the weedy instinct to produce so abundantly that its self-seeding drives out more retiring plants.

It is closely related to many fabulous hybrids created in the Orient, Europe, and America. The best known to modern gardeners was developed in this century by American horticulturist Luther Burbank, who produced so many new and improved varieties of plants. He decided to commercialize the oxeye's stamina and irresistible appeal. After years of experimenting, he combined the most promising characteristics of the oxeye daisy, the Nippon daisy, the maximum or moon daisy, and the German daisy to produce the Shasta daisy, which he named for the majestic mountain rising above his experimental nursery in California.

Ancient Aristocrat of the Orient

Even in the oldest literature of China the chrysanthemum appears as a sophisticated flower. The history of its development from a mere daisy is lost in the shadows of long ago, for behind the elegant blossoms seen by the first travelers from the West lay centuries of care and purposeful development.

Confucius wrote of the yellow chrysanthemum 500 years before Christ was born. A thousand years later T'ao Ming-yang, a Chinese scholar and botanist, became enamored of chrysanthemums and did much toward developing a new species. The fame of his chrysanthemums drew many pilgrims to the city in which he dwelt, for in old China, as in Persia, exquisite flowers were thought to be a focus for meditation and devotion. So popular were his gardens that the city became known as "chuh-sien," "the City of Chrysanthemums."

The oldest Chinese books extant refer to thirty varieties of chrysanthemums under cultivation at this period. China watched over them with a possessive eye, and produced the parents of our modern hybrids, *Chrysanthemum morifolium* and *C. indicum*, whose progeny of hybrid species have spread throughout the gardens of the temperate world. But all, even the great globular types, have their source in the wild daisy that grew in China and probably looked not unlike the immigrant oxeye daisy of our summer fields and roadsides.

Japan

There is an old legend that says the Empire of Japan was born when a shipload of twelve maidens and twelve youths of the Chinese nobility set out on a mission to find the "Herb of Youth," in order to sustain the life of their beloved emperor. They carried with them bamboo baskets of golden chrysanthemums, China's most precious product, to trade for the herb.

After a storm-tossed voyage of many weeks they reached one of the islands of that archipelago, only to find it uninhabited. Seeing that their battered ship could carry them no further, they planted their chrysanthemums and settled down to build an empire. And that is why the flag of Japan bears a golden chrysanthemum, which many Westerners misinterpret as the rising sun, but which in reality represents the sixteen-petaled chrysanthemum, the emblem of the mikado.

Behind all legends lies some historical fact lost in the mists of the ages. Perhaps this legend symbolizes the recognition of China as the birthplace of chrysanthemums, and the tradition that it was the exclusive possession of the Japanese aristocrat. Even a century ago its cultivation, actually forbidden to common folk, was an occupation for the dilettante nobility, who continued to develop fantastic new forms and colors. The viewing of chrysanthemum displays was a favorite recreation for the elite, and attendance at the Imperial Chrysanthemum Gardens a social ambition. So the chrysanthemum became a status symbol, and in 1876 the emperor created the Order of the Chrysanthemum, the highest honor he could bestow.

But World War II changed all that. Today the emperor is an ordinary mortal, and the chrysanthemum is a flower for all the people, who are free to view, to buy, and to grow it—a symbol of the new Japan, which is changing at a furious rate.

Although the new freedom has encouraged commercial growers to apply their ingenuity in cultivating new forms of pyramid, cascade, bonsai, and topiary chrysanthemums, Japan has also retained her priceless chrysanthemum traditions. Latter-day research reveals that chrysanthemums first reached Japan as seeds from Korea in the fourth century, and eventually were reinforced by new forms from China. The chrysanthemum became the national flower of Japan in the year 910, at the time of the first Imperial Chrysanthemum Show.

Today's Chrysanthemum Festival manifests the time and effort Japan has dedicated to chrysanthemums. What other flower has commanded a whole nation's enduring devotion?

Chrysanthemums in Europe

The Dutch, who were among the first Europeans to have access to the Orient, besides being aggressive sailors and astute merchants possessed an agricultural eye, and they probably brought the first chrysanthemums to Europe. Records show

that a half-dozen varieties were grown in Holland in the eighteenth century, but, strangely, Europe in general showed
little interest in them. Only the French, among whom the
Huguenots had a tradition of horticulture, found the chrysanthemum worth developing. They imported a variety of
forms, and many chrysanthemum gardens throve in the south
of France, where the climate was conducive to the production
of seeds for commercial purposes. It was here that Old Purple,
one of the basic chrysanthemums of European gardens, began
its career and helped to establish the chrysanthemum as a
garden flower.

England

England was slow to appreciate the chrysanthemum, but
nowhere else in Europe today do individual growers and
exhibitors show such enthusiasm for it.

In 1764 Phillip Miller, of the Chelsea Gardens, records
the arrival of the yellow Chinese chrysanthemum, and its
early death. Old Purple arrived at the Royal Botanic Gardens
at Kew in 1793, but it, too, drew little attention. In 1815 the
magazine *Botanical Register* published illustrations of the new
decorative varieties of chrysanthemums being grown on the
Continent, but the English continued to display an almost
total lack of interest.

Robert Fortune, England's greatest plant hunter,
brought the Chusan daisy from China in 1846. At the time,
England ignored it, too, but its small rounded blossoms
became popular in France, where they were developed and
named "pompons" because they suggested the wool pompons
on the French soldiers' hats. In 1861 Fortune brought back
the large, shaggy hybrid Japanese chrysanthemums, but English gardeners disdained them. And although the pompon
chrysanthemum presently became popular there, England did
not turn into a nation of enthusiasts until this century, when
the perennial garden chrysanthemums were established.

Chrysanthemums in America

The first chrysanthemums to reach America arrived in 1798 when John Stevens, a nurseryman in Hoboken, New Jersey, imported the durable Old Purple. By 1850 the Chusan daisy had been imported, and proved immediately appealing. William Prince, of the Prince Nursery on Long Island, commented on a growing interest, in his *Short Treatise on Horticulture*, which lists forty varieties of chrysanthemums available for American gardens. Serious breeding was clearly under way in the United States.

But this, too, was interrupted by the Civil War and half a century of intense pioneering. Only on a few estates and in large nurseries of the East Coast was chrysanthemum cultivation carried on, and even there it was limited to the large varieties and the pompons, grown principally for conservatories or the cut-flower trade.

By 1900, however, a few American nurserymen were producing their own seed, and imports of the newest species stimulated a growing demand. The Chrysanthemum Society of America was founded, and held its first exhibit in 1902—in Chicago, which indicates that an interest in gardening was moving westward.

But few chrysanthemums were adaptable to sections of the country where severe winters and torrid summers prevail. In 1909, L. H. Bailey wrote in *The Standard Cyclopedia of Horticulture* that only 10% of available chrysanthemums were suitable for planting in most areas. Nearly all the loveliest varieties of this Asian aristocrat still belonged in the conservatory.

Two men stand out among the devoted workers who have transformed the hothouse beauties into America's most popular fall garden flower. The first is C. Elmer Smith, who introduced 400 varieties and did much to lay the basis for the commercial success of large-flowered species with staying power for shipment and display—and football games. The second is Alex Cumming, who introduced the Korean hybrids, with their six-inch flowers and incurved rays, to American gardens in 1932. His work continues through his son, Roderick W. Cumming, author of *The Chrysanthemum Book*, published in 1946.

Today's catalogues testify to Smith's and Cumming's success. We are offered more than 150 species of chrysanthemums—from China, Japan, Korea, Australia, England, France, and Holland—and give them all a hearty welcome. The imperial chrysanthemum from Oriental palaces lends its profuse beauty to millions of small American gardens, demonstrating again the democracy of nature.

DANDELION

Today there are more dandelions in the United States than there are dollars in our national debt. But there were no dandelions at all when the white man came. How did they get here? Where did they come from?

Origin and History

The dandelion is a plant of the Temperate Zones that probably originated in Asia Minor, but it had spread throughout the known world before written history. In the East, where the Chinese call it "earth nail," its long taproot and green leaves have been used for food and medicine since antiquity. In Japan, where it has been grown as a decorative plant, 200 varieties in white, orange, black, and copper are cultivated by florists, and a national Dandelion Society existed until the twentieth century.

In the Mediterranean world the Greeks learned early to tenderize the leaves by bleaching, and ancient myths tell that Hecate, the goddess of earth, moon, and the underworld, honored Theseus with a salad of dandelion greens after he had slain the Minotaur. In Rome, too, the dandelion was commonly used in salads and stewpots, and when the Roman legions invaded the Rhineland and Gaul they rejoiced to find it growing there.

In Britain, Caesar found that the Celts claimed the dandelion as truly their own, a source not only of food but also of a wine that they made by fermenting its golden flowers. Dandelion wine is the most highly regarded of England's country wines to this day.

The Anglo-Saxon tribes who settled in the British Isles after the Romans withdrew used the dandelion to control scurvy and as a laxative and diuretic. It became one of the medicinal herbs planted in the physic gardens of monasteries, and its botanical name, *Taraxacum officinale,* shows that it was registered in early monastic hospitals as a useful medicinal. In France this familiar pot herb was universally used, and the Normans, who conquered England in 1066, called it "dent-de-lion" because its sharply indented leaves suggested the teeth of a lion. The Saxon serfs soon corrupted the name to "dandelion," and so it has remained.

The French concocted a gourmet dish, still a popular delicacy, from dandelions—a spring dessert made by dipping flower heads into batter and frying them in deep fat. But the Elizabethan herbalists give little space to the dandelion, seeming to accept its universal use. A comment by John Evelyn, a diarist of Samuel Pepys's time, is typical; he notes that it is "a cleaning spring green."

Dandelions in America

It is obvious that by the time the Puritans set out for New England the dandelion was considered an essential plant for food and health. Nevertheless it is not on the long list of seeds ordered by Governor Winthrop for Boston, nor does Governor Bradford mention it in his accounts of Plymouth

Colony. The earliest self-appointed chronicler of Salem's gardens, Francis Higgins, ignores dandelions, and so does William Wood in his detailed account of the settlers' horticulture, *New Englands Prospect*, which he wrote in 1634.

There was a reason for this. The dandelion was not an economic plant but a common green for the stewpot, a "dosing herb," and so its seeds were among those each woman was expected to take with her for the family garden plot. We know that the dandelion did grow in those first gardens, protected from rabbits, woodchucks, and deer by fences woven from long branches of willow and plastered with mud. Elizabeth Winthrop, a relative of the governor, who had been trained in medicinal herbs by her father, a London apothecary, planted her physic garden, like most women of the colony, with dandelion, scurvy grass, marigolds, and other dependable dosing herbs from England.

As European colonists settled along the Atlantic coast they continued to bring the dandelion for their gardens—the Dutch in New Netherland, the Germans in Pennsylvania, the French on the Saint Lawrence, and the English in their southern colonies. When the forests were cut down, the dandelion escaped to the cleared fields and roadsides, and soon grew so abundantly that it was no longer given garden room. It followed the lumbermen over the mountains and ran rife across the prairies. The Apache Indians welcomed it, and are said to have scoured the plains for it days before their spring feasts. Thus the dandelion preceded the pioneers on their westward migration, and was waiting for them when they reached the new land, offering a food, a tonic, and a heady wine.

During the Civil War the dandelion was used by both the north and the south as a medicinal herb, and southerners, who had been cut off from imported foods by blockades, used its dried ground roots as a substitute for coffee. Indeed, they went on doing so long after the war.

Modern Military Use

During World War I, health departments both here and abroad publicized the high food value of dandelions—a green, delicious food, forgotten by urban populations but growing abundantly in the wild. By the time World War II began, the English had virtually stopped growing herbs, because it was cheaper to import them. But when the war cut England off from her Far Eastern and European sources, the Ministry of Health organized groups of women to search for native herbs. The dandelion was invaluable. Its roots, which can be dug from September to March, were used for a tonic, a laxative, and, once more, for a caffeine-free coffee, and its leaves provided spring salads and a cooked vegetable. It was ultimately honored for its war service by being given a place in the British Pharmacopoeia.

In Russia the dandelion not only provided food during the war, but also proved to be a source of latex. Russia has since spared no expense to convert the dandelion's latex, which is found in all parts of the plant, into commercial rubber.

Varied Modern Service

The dandelion now girdles the Temperate Zones of the earth. It has adapted to modern man and offers services for several of his present needs.

It is, for instance, an aid to the conservation of wildlife. Our exploding population and the proliferation of industry, housing, and superhighways have destroyed millions of acres of wildflowers. Few cultivated crops are planted near urban and industrial areas, and over wide areas the wild bees are dying of starvation. But the dandelion, which will grow anywhere—on disturbed ground, wasteland, slag heaps, even on condominium lawns—has come to their rescue. It is a rich source of food for them and is particularly valuable for young bees in the hive, which require a constant supply of that mixture of pollen and honey known as bee bread in order to mature and be ready to pollinate the late summer flowers. Without these bees the remaining wildflowers, fruits, and berries would perish, too.

Plant scientists are turning to the dandelion for help in increasing world food production. Research has shown that its blossoms exude ethylene gas at sunset, and experiments are being carried on with mass plantings of dandelions in orchards. The results have been amazing. The gas causes the fruit to ripen days earlier, and increases market profits with bigger and better fruit; the dandelions themselves have grown larger. The scientific name for this mutually beneficial association is "symbiosis," and its economic possibilities are extensive.

Beauty, Utility, Laughter

The dandelion has a beautiful golden flower head whose tiny blossoms, each with a pistil and stamen, soon turn to fluffy parachutes tipped by dark seeds—35,000 of them to the ounce—that are carried far and wide by the wind, achieving maximum distribution and incidentally providing food for wild birds. To those who would exclude the dandelion from their lawns, Louis Bromfield, best-selling novelist turned ardent Ohio farmer, pointed out that its long taproot is nature's way of aerating the ground for the short grass roots, thereby encouraging greener lawns and greater beauty. It also brings up from the depths elements not available to shorter-rooting plants; so dandelions are a valuable addition to your compost heap.

Every part of this ancient herb is useful in our modern world—its leaves, its blossoms, its roots, the latex in its veins, the very air it exhales. Finally, it offers our children a legacy of laughter. Generations have picked bouquets of its golden flowers, spent happy hours twisting curls from its stems, and told time on dandelion clocks by blowing at its fluffy seed heads. Perhaps the dandelion's most delightful gift to humankind is centuries of laughter.

DAY LILY

Gay, robust, irresistible, day lilies are the best known and best loved of all the lily tribe, all around the world. For centuries they have flung their gold and incense across gardens, fields, and roadsides.

Their story begins in the dawn of history, in central China, where fields of day lilies were cultivated for food. The people ate the flower buds as a spring tonic and a summer delicacy, and surplus buds were gathered and dried for use in winter as both food and medicine. It was believed that a powder made from the buds could dispel grief, relieve pain, purify the kidneys, and favor the birth of a son if it was worn in the girdle of one's gown during pregnancy. And the clumps of green leaves provided a forage that cattle ate with apparent relish.

So valued was the day lily as a medicine and food that it was carried, along with rhubarb, in the caravans that followed the ancient silk routes all the way to eastern Europe, where these coveted healing plants were worth their weight in gold.

Though the day lily early established itself in Siberia, Turkey, and Hungary, it did not reach Western Europe until 1500. By that time its medicinal and food values were forgotten. Around 1595 it was introduced into England as a rare garden flower under the name "liricom fancy," arriving by way of Holland with the waves of Huguenot refugees fleeing France.

John Gerard notes with pride: ". . . these Lillies do grow in my garden as also in the gardens of Herbalists and lovers of fine and rare plants." He adds, "It is fitly called Faire and beautiful for a day and so we in English may rightly terme it the Day Lillie."

Day Lily in America

So admired was the day lily in England that it was among the first flowers brought to the colonies. By 1695 both the tawny day lily, *Hemerocallis fulva*, and the lemon lily, *H. flava*, were grown in the dooryards of New England and the walled gardens of Virginia. Its adaptability was recognized by the Dutch colonists, who planted it in the early gardens of Manhattan, and by the Quakers, who brought it over for the first gardens on the Delaware. And, like the lilac, it became a dooryard favorite in the age of pioneering. Its endurance was a promise of the future; its abundant blooms were a present pleasure.

The *Hemerocallis* is a true member of the lily family, although it grows from a fibrous root—a fact that aided its escape from old-time gardens. It followed the settlers across the Alleghenies and the plains, and spread in masses to fields and glades and roadsides. Today a clump of day lilies is often the only remaining trace of a forgotten graveyard or an abandoned farm—nostalgic testimony to some pioneer's dream.

In our present-day conservation programs, the day lily should be recognized as a plant with practical as well as ornamental characteristics. Its masses of roots and dense clumps of leaves make it a perfect fortifier for banks where road building has scarred the earth, and it serves as a splendid ground cover even in partially shaded or damp locations over a wide range of states.

It is one of the delights of the summer garden. Its several varieties may bloom from June to September, and its leaves are always green and fresh. Though its prodigal vigor and its tendency to spread reduced its popularity early in this century, its beauty fascinated hybridizers and nurserymen. The New York Botanical Garden has a bevy of growers who have created day lilies in a wide range of shades from deepest red to palest pink. The number of its admirers steadily grows, and they have founded the Hemerocallis Society, which boasts some 2,000 members.

Culinary Research

If you wish to indulge in research in the gastronomic and medicinal properties of this lovely flower, here is an old recipe: "Pick the blossoms near dawn as they open. Remove the necks where they join the stem. Bake in a low dish with rich milk, butter and salt." And if you should suffer a mishap in the baking, take comfort from Gerard, who recommends that the roots and leaves of the day lily "may be laid with good success upon burnings and scaldings."

EUCALYPTUS

In 1688 a pirate ship that had floundered in the South Seas for weeks dropped anchor in a harbor on the southeast coast of Australia. While the thirst-ridden crew dug for water on the parched tropic shore, the captain took his notebook and wandered inland to examine some giant, fragrant trees in the distance. Thus was the eucalyptus, perhaps the world's most versatile and valuable tree, discovered by one of history's most controversial adventurers, the first Englishman to reach Australia.

William Dampier was no ordinary buccaneer. He was an author and natural scientist of repute, recognized by such men as John Evelyn and Samuel Pepys; he was consulted as an expert hydrographer and navigator, and was appointed by the British Admiralty to be naval officer in command of a later expedition to the South Seas. His botanical sketches, notes, and herbarium were welcomed by the Oxford Botanic Garden, where they may be seen to this day.

In 1697, some time after his return to England, he published a book about his adventures, *A New Voyage Round the World*, in which he describes the giant mysterious trees of Australia and notes that they were fragrant with gummy exudations that fell to the ground like manna and were eaten by the aborigines. He tells us, incidentally, of his discovery of the castor-oil bean on the Amazon, and of the breadfruit tree that would make world history and bring Captain Bligh, of *Bounty* fame, to Australia as its governor.

Dampier's account of Australia aroused such excitement about the unknown island continent that further explorations were immediately planned. But a series of wars, and the priorities they imposed on the British, postponed the execution of the plans for nearly a century.

Captain Cook and the Eucalyptus

Seventeen-sixty-eight was the year of the transit of Venus, the rare phenomenon when that planet passes between the sun and the earth. It is important to science because the distance from the earth to the sun can be measured by the time it takes Venus, as observed from the earth, to cross from one side of the sun to the other—a magical event that will not occur again until 2004.

The Royal Society of London, England's oldest scientific organization, decided to finance an expedition to Tahiti, from which the transit of Venus could best be observed. To head it they chose James Cook, a young officer in the royal navy who had just returned from a voyage during which he distinguished himself by surveying and charting the Saint Lawrence Channel and the coasts of Labrador and Newfoundland.

Joseph Banks, a wealthy young Englishman who possessed the intense interest in natural history and the fascination with unknown places that were so common in the eighteenth century, accompanied Cook on the expedition for the purpose of collecting specimens of plants and animals, and exploring the South Pacific as opportunity offered. He chose Daniel Solander, a protégé of Linnaeus, as botanist for the expedition.

After leaving Tahiti, where the transit of Venus had duly been observed, the expedition sailed to New Zealand. Cook mapped the coast before sailing on to map the eastern coast of Australia, and eventually reached Botany Bay, so named by him and Banks because of its varied vegetation, which differed dramatically from Europe's.

Banks and Solander, filled with wonder, found Dampier's giant trees; but the sweet gum that seemed to exude from the leaves was really, they learned, the result of a scale that a tropical insect deposited on buds and twigs. Besides eating this manna, the natives used the tree itself as a source of food and drink. They would break pieces from its great roots, which in this arid land were turgid with stored water, and suck at them to quench their thirst. The dried root itself they pounded to flour for bread; and the tree's seeds, too, provided a nourishing food.

Solander and Banks collected specimens and seeds to take home to Europe. Strangely enough, their notes on this three-year expedition, which eventually circumnavigated the globe, were not published until 1900. But tales of the South Seas and of the marvelous tree circulated just the same, and stirred the interest of other European nations.

Captain Cook, meanwhile, set out the following year to explore the South Seas further. At the last minute Banks withdrew from the ship's company, and Cook accepted in his stead two German naturalists, John Reinhold Foster and his son, Johann Georg, who, as things turned out, played only a small role in the history of the eucalyptus.

It was on this voyage that Cook, a notably enlightened man for his day, established the strict dietary and hygienic regulations that prevented scurvy and thus dramatically reduced the crew's death rate. Whereas 118 men had died during the first voyage, there was only one death during the second, which also lasted three years—a record that so impressed the government that Cook's regimen was made a standard for the British navy.

Cook was prevailed upon to undertake yet another voyage, this time to map the west coast of North America and to search for the Northwest Passage. The search of course proved fruitless, but he enjoyed a greater success: the rediscovery of the Sandwich Islands, now Hawaii. David Nelson, an able botanist from Kew Gardens, accompanied Cook on this voyage, and four years later took back to England the first living specimens of the Australian eucalyptus, which were planted at Kew.

In 1789 Charles L'Heritier, a renowned French botanist, wrote the first detailed description of eucalyptus, and gave it its botanical name. This is a combination of two Greek words: *eu*, which means "well," and *kalyptos*, which means "covered." It refers not to the shade the tree casts but to the lid or operculum that covers the flower bud until it is thrown off by the opening blossom, a characteristic that distinguishes the eucalyptus from all other Australian plants.

Essence of Australia's Diversity

There are about 300 species of eucalyptus, and they are strikingly different from one another. They belong to the myrtle, Myrtaceae, family, which also gives us cloves, pimiento, and allspice. They all have entire leathery leaves, which are asymmetrical and therefore able to turn their narrow edge to the sun and reduce evaporation, and they all contain a fragrant volatile oil.

The eucalyptus is the tallest known tree, even taller than the Sequoia. The karri, a white gum, which grows straight toward heaven like a great candle, averages 300 feet, and may reach 180 feet before its first branch. The trunks are often seven or eight feet in diameter, and sometimes they are buttressed by supporting roots, remaining supple and yielding despite their enormous size before the fury of tropical gales.

The earliest Europeans in Australia soon discovered that, for timber, the eucalyptus was comparable to the best oak and maple. But its growth was far more rapid than that of European hardwoods, and the Western nations soon rivaled one another in planting eucalyptus at home and introducing it to their colonies.

Eucalyptus in Europe

The first experimental planting of eucalyptus in Europe, spurred on by L'Heritier's glowing accounts of its economic value and its beauty in landscaping, took place in France at the end of the eighteenth century. By 1804 the Empress Josephine had a eucalyptus tree planted in her garden at Malmaison, where it was regarded as a great curiosity.

The Portuguese watched with interest the rapid growth of experimental plantings in France, and undertook plantings of their own along their eroded coast, where the trees had to withstand long periods of rain. They flourished, and many towns began planting eucalyptus as shade trees along streets and highways—a practice that was soon enthusiastically followed by other Mediterranean countries, until eucalyptus trees came to seem native to all of southern Europe.

The beauty, variety, and intricate design of eucalyptus seeds soon led to their use in the manufacture of rosaries throughout Portugal and in Spain and the missions of Spanish America. Rosaries of eucalyptus seeds strung on cowhide are exhibited today in the Indian museums of the old missions of southern California, and many are of great beauty.

In Italy eucalyptus was planted in the Roman Campagna, whose malarial swamps were notorious, and transformed the area with its beauty and fragrance. Periodic plantings of

eucalyptus on hillsides denuded by ancient abuse have provided Italy with essential lumber. Eucalyptus has withstood,
as well, the bitter frosts of India and the heat of tropic Africa;
but to the mystified frustration of the English, it refuses to
accept the relatively mild winters of the British Isles.

Eucalyptus in Africa

The eucalyptus was introduced to Africa by European
colonists. In Kenya they cut down some native virgin forests
to make way for eucalyptus trees, which would produce repeated crops of profitable lumber. In Pretoria modern tourist
camps are constructed of eucalyptus poles and thatched with
its leaves. There are avenues of magnificent eucalyptus trees
in Algiers, and whole forests of them in the lake regions.

The success of eucalyptus throughout Africa has led to
its being planted even in the Sahara, where the sands are re-
peatedly sprayed with chemicals to keep them from blowing
until the young trees are established. The experiment, begun
in 1961, has been so successful that the scientists in charge
believe the fabulous tree may eventually transform the desert
into a productive forest.

Eucalyptus in South America

The eucalyptus was introduced to Peru by the Spanish padres
soon after Cook's South Seas voyages. These inveterate experimental horticulturists must be credited with the many old
groves of immense trees that now seem an integral part of the
landscape and are of great economic value.

The government of modern Peru has taken a leaf from
the padres' book, and eucalyptus is playing a principal role in
an AID program, in which the United States is co-operating
with the Forestry Service of Peru. Already 8,000 acres of
nonproductive barren land on the Andean slopes have been
planted with millions of eucalyptus seedlings. The goal in

view is a forest of 250,000 acres that will stabilize the soil, diminish silting of streams, and provide a valuable timber crop.

Torkel Holsoe, professor of forestry at the University of West Virginia, headed the Peruvian project. Peace Corps volunteers who are graduate foresters give technical aid, and Peruvian workers receive part of their pay through the Food for Peace program. The seedlings are cultivated in Peruvian nurseries and sent to the mountains in plastic bags of compost. In six years their trunks will reach five inches in diameter, and the first cutting and thinning will take place, the young timber being sold for fence posts, mine props, roofing, pulpwood, and fuel, and returning enough to pay for the initial investment. New growth, sprouting from the stumps, may, if left uncut, grow into giant trees.

The eucalyptus was introduced into Chile in 1823, when the universal blue gum, *Eucalyptus globulus*, was planted there. Now many species of eucalyptus grow in the long Pacific state, with its wide range of climates, shading the streets of its cities and massing in forests, upon which an extensive lumber industry is based.

The eucalyptus finds an ideal home in Colombia, and grows there to immense heights. Great beams of its superb wood are featured in homes and public buildings, and it is widely used for furniture, because it has proved immune to tropic heat and insects, and will not rot or warp in the rainy season. The same characteristics make it an excellent timber for boats and docks.

In Brazil eucalyptus forests are something of an anomaly, for Brazil has hundreds of thousands of acres of virgin oak and maple, with trunks seven to eight feet in diameter, that have never been cut because they lie beyond the reach of transportation. It is cheaper to plant eucalyptus trees in accessible areas, where they provide abundant timber and resins for commercial and medicinal oils.

Eucalyptus in China

The Chinese consul to Italy in 1894 marveled at the avenues of eucalyptus along the city streets. When he learned of their versatility, their hard wood, their rapid primary, secondary, and tertiary growth, he determined to introduce them to China. The first experimental plantings were made in Canton, Foochow, and Hong Kong, where shade trees were badly needed. But the Chinese showed little interest and called them "foreign, stinking trees."

Although a few nurseries were established by 1926 and there was some limited experimental planting, eucalyptus was little used until 1949, when China came under the influence of Russia and hundreds of Communist advisers attempted to bring China's farming and forestry up to date. Eucalyptus forests proved adaptable in southern China and are now intensively cultivated there. Miles of eucalyptus are planted along railroads, streets, and highways, and planned forests provide lumber for everything from furniture to firewood.

Eucalyptus in the United States

Eucalyptus was introduced from Australia in 1853 by the rapidly growing new state of California. It had been grown in Spanish mission gardens throughout the American southwest, which had so recently belonged to Spain, but little interest was shown in it until after the Civil War, when the rush for free land began. A great expenditure of energy and finance went into the founding of towns in river valleys and along the Gulf Coast during the land boom, but malaria soon invaded these settlements, and proved so devastating that many of them were abandoned. It was then observed that Mexican towns in these localities were thriving, and were untroubled by malaria or mosquitoes. The secret seemed to be the eucalyptus trees that the townspeople planted along their rivers. American settlers began to adopt the practice. The fast-growing eucalyptus soon provided blessed shade and windbreaks; swamps dried up, malaria was reduced, and the fragrant leaves seemed to discourage mosquitoes. Pioneer towns in the river bottoms then flourished.

But it was not until 1901, when eucalyptus trees dotted the landscape of California, Arizona, and the Gulf states, that the relation among mosquitoes, malaria, and the trees was generally understood. During the Spanish-American War, malaria and yellow fever had killed thousands of Americans in our southern ports and the West Indies. Intensive research into the cause and control of these diseases revealed that both were carried by the mosquito, and that the mosquito could be controlled by drying up the swamps where it bred. Research also proved this could be done naturally and permanently by planting the appropriate species of eucalyptus near wetlands; the great roots of a single tree could demand as much as 300 gallons of water for routine use.

The then newly founded Arizona State Experiment Station, which conducted much of the research on eucalyptus, has continued its interest in the tree to this day. In California, where it is much grown, large commercial factories once did a profitable business in eucalyptus resins and medicinal oils. And today's visitors are deeply grateful for the landscaping of the high ground above the ocean, where the lower-growing, pink-blooming crimson eucalyptus, *E. ficifolia*, enlivens miles of highway with sheer beauty.

Though more than fifty species of eucalyptus are now grown in the warmer sections of the United States, 80% of the trees are blue gum, which serves the widest variety of landscape and timber needs. New uses for it are found in the commercial production of fiberboard, pulp paper, cordage, thatch, and perfume. The fragrance and beauty of eucalyptus have even penetrated to towns and cities of the northeast, if only in the form of the ornamental sprays sold in florist shops. The mature, varied leaves of the eucalyptus grow to twelve inches, but the young shoots bear silvery-blue, opposite, roundish leaves that are most beautiful and decorative. The twisted stalks of *E. polyanthemos* are particularly popular, and there is a growing demand for the vast quantities shipped from the West Coast each year.

How can one describe the King of Botany Bay? Because of its grandeur, its varied ways of growing and flowering and seeding, and its beauty and service to man, one's instinct is to bow before it in gratitude for its manifold gifts.

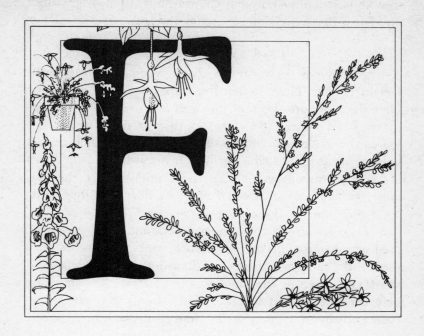

FLAX

Flax has served man longer than any plant that is not a food: it was sown, reaped, cured, spun, and woven into cloth before history was written.

It is native to Eurasia, including the Mediterranean area, the cradle of Western man. We know from archeological remains that linen was made in prehistoric Mesopotamia, and by the lake dwellers of neolithic Switzerland, and as far west as the Pyrenees. It was worn as a symbol of purity by Greek and Hebrew and Egyptian priests, and in Egyptian tombs the mummies of Pharaohs lay wrapped in fine linen bindings for thousands of years. Detailed wall paintings in the tombs represent the history of flax in the Nile valley, and show us that its basic cultivation and the transformation of the fiber into fabric has changed little through the ages.

From Field to Fashion

A field of flax is ready for the reaper when it averages three feet in height. The cut stems are first soaked in water for some days, or spread out in the open and exposed to the elements for a matter of weeks, so that bacterial action can loosen the outer bark and dissolve the gummy substance that holds it to the inner stem. The bark is discarded and the stems are dried, crushed, and beaten: once upon a time this was done with broadswords and flails. The fibers are then separated and spun into threads for weaving.

They are strong and sinuous fibers, and they were once used with hemp to make stout durable ropes. Such ropes were used in the pulleys that lifted and swung the great blocks of stone into place in the building of the Pyramids; and long before Rome learned to make cloth from the fibers, sailors twisted them into ropes for the ships that sailed the Mediterranean and Adriatic seas.

In Mesopotamia and Egypt, linen from flax became the chief source of clothing in very early times. Herodotus, writing in 455 B.C., says the priests of Egypt were garbed in fine linen, and the Book of Exodus tells us that Solomon bought linen from Egypt and that it was in raiment of fine linen that Pharoah clothed Joseph. Through many centuries—in the Mediterranean world, in the spreading Roman Empire, and in the European nations of the Middle Ages—linen remained a measure of wealth.

Paper for Western Man

As early as 2200 B.C. the Egyptians had learned to make paper from strips of papyrus pith which they laminated together with the adhesive sap of the plant. These strips were then dried, pounded, and pressed into sheets that could be written on. Papyrus was used until A.D. 400 and was the forerunner of parchment—shaved sheepskin—which served as writing paper in Europe until the twelfth century. Meanwhile, China had developed a paper made of bamboo and mulberry fibers in A.D. 105, and this technique spread to Japan, Korea, and throughout the Far East. Then, about A.D. 750, a superior paper was created of flax fibers in Samarkand, still part of the Moslem Empire.

In the next five centuries the Moors built paper mills and introduced flax from Cairo to Morocco and Sicily. In 1144, they erected a paper mill at Xátiva, Spain, and flax flourished there.

During the Middle Ages the Crusaders invaded the Moslem strongholds and brought back to Europe, among other things, the art of papermaking from flax fibers. Paper mills were opened in France and Italy after the Second Crusade.

The parchment makers' guilds, which were very strong in some localities, loudly protested the manufacture of linen paper, which destroyed their livelihood and their political power. They burned the flax crops, destroyed the mills, and won a temporary victory by having paper banned in many areas. But history and economics worked against them. The introduction of mechanical printing about the middle of the fifteenth century—generally attributed to Johann Gutenberg—increased the demand for paper. The production of flax could not keep up with it. Dozens of flax substitutes were experimented with: thistles, rushes, and even asparagus—but all proved unsatisfactory. Then the French naturalist René de Réaumur, who had observed how the paper wasp built its nest, suggested that paper might be manufactured from wood pulp. By 1800, mills were producing wood-based paper in France, and the practice soon became universal.

Flax in America

No flax was found on the Atlantic coast, but it was introduced to all the early colonies as essential for clothing and domestic linens. Its use, however, was quickly curtailed because of cotton, which is native to America. When Columbus arrived in the Caribbean, he found the natives wearing clothes of fine cotton, which he offered as proof that he had reached the Indies, known in Europe as its source. The Jamestown colony was planting cotton by 1619, and in a few decades it was grown throughout the south, establishing a belt that exported cotton to the textile mills in our northern states and Europe. Except for a short attempt to revive flax during the Civil War, when the north was cut off from the cotton states, linen has never been of economic importance in America.

Flax fiber, however, was essential to our own early paper mills. In 1690 William Rittenhouse, a Mennonite bishop interested in the economic success of his community, built the first paper mill at Germantown, near Philadelphia. His mill, and all the other paper mills built in colonial times, used flax fiber and linen rags; but the new and rapidly growing nation born of the American Revolution demanded more paper than such mills could supply. In 1817 John Gilpin opened the first wood-pulp paper mill on the Delaware, and linen paper became an expensive rarity.

Flax nevertheless grew steadily in economic importance, because flaxseed and linseed oil are essential to numerous products.

Flaxseed has been used in medicine since antiquity, for liniments, cough syrups, and salves with which to treat boils and similar infections. An early book on household health directs the mistress to stamp flaxseed, mix with cucumber root or boil with mallow or chickweed or violet leaves, and mix well with hot hog grease "to affect and relieve all manner of paine." And flaxseed has long been listed in the United States Pharmacopoeia for its medicinal value.

Increasing amounts of flaxseed were also required for linseed oil, a drying agent in commercial paints and varnish, and in the oil paints used by artists. Gerard sums it all up neatly: "Flax seed is profitable for physicke and Surgerie, for paints, for picture artists, and other artificers."

American ingenuity soon found even more important uses for flaxseed. The residue left when the oil has been pressed out is made into flaxseed cakes, a convenient and nutritious way to fatten cattle. A still wider use occurs in the manufacture of oilcloth and linoleum, whose surface is made from oxidized linseed oil mixed with ground cork and various other ingredients.

In the United States today, flax is grown principally in the northwestern states. It is more demanding than any other commercial crop, and must have rich, well-drained land and a damp subsoil. It is often grown as a two-product crop, for its seed and its fiber, but the finest linen is produced from stems cut before the seeds are ripe. On poorer land it is grown for the seed only. In Belgium and Ireland, which have produced the world's finest linen, the land is continually and heavily fertilized.

Botanical Bits

Flax is a hardy annual. There are more than a hundred species scattered over the globe, but only linen flax is of great commercial importance. Linnaeus aptly called it *Linum usitatissimum*, which means "most useful of the linen family." The word "flax" is universal, and is derived from the process of flailing or flaying the fibers.

The seed is sown in early spring. The blue five-petaled flowers bloom in June and July at the end of slender branches; the leaves are slight and lanceolate, the seed pods dark smooth knobs.

Perennial flax is an English garden flower that has been cultivated since the sixteenth century. Varieties of perennial flax, which were introduced to England from the Continent and Algeria and whose flowers come in shades of yellow, blue, pink, and white, have been brought to America. Native

forms of perennial flax have also been found in the Appalachians and on our West Coast.

New Zealand Flax

New Zealand flax, *Phormium tenax*, which several pioneer
plant hunters found in the Antipodes, is cultivated in a limited
way in California. It is not a flax, however, but a member of
the lily family. Its six-foot irislike leaves produce excellent
fibers that are used in the manufacture of mats, cloth, rope,
and similar products. But it cannot compete with the fine lace,
linen, and paper of flax fibers, or with the commercial products of flax seed.

FORGET-ME-NOT

The forget-me-not is seldom seen among the handsome hybrids in our modern gardens, but the early colonists brought
it to America, and it was treasured until a few generations
ago. Its frivolous beauty was tolerated by the Puritans because, under the name "scorpion grass," it offered a cure for
the stings of scorpions and the bites of adders and other
venomous creatures that might inhabit the dark forests beyond the settlement.

The forget-me-not's appeal was enhanced by the familiar
romantic legend in which a knight in shining armor, riding to
battle, stopped to tell his ladylove farewell. Thoughtlessly,
she urged him to gather for her a parting bouquet of blue
flowers from the crag above the lake. He slipped and fell, but
as he plunged to his death he tossed her the bouquet, crying,
"Forget me not."

Variations of the legend are doubtless told in many parts
of the world, for species of forget-me-not are native throughout Eurasia. Perhaps the Crusaders carried a version to the

Holy Land, whose troubadours sang of forget-me-nots as nostalgic reminders of faithful love and walled gardens at home. And it was used as late as 1465 by the ladies of Europe's courts, who thought it a sophisticated flower symbolically connected with subtle affairs of the heart.

Its tradition as a symbol of fidelity blossomed again in the "language of flowers" during the Victorian era. In this romantic concept, each flower was said to have a meaning, and young men carried on clandestine flirtations by sending bouquets that could be translated into messages of love. In such messages, the forget-me-not, to indicate undying devotion, was invariably included.

Mundane Service

The forget-me-not also had an ancient and creditable history as a healing herb. Dioscorides in the first century recommended its leaves as a cure for a scorpion's sting; he affirmed, too, that a decoction of the leaves boiled in wine would prevail against the bite of an adder or a venomous beast, and that a healing ointment could be made by mixing the leaves with oil and wax.

We learn from Gerard that what came to be called "scorpion grass," *Myosotis scorpioides,* did not originally grow wild in England. He was, he tells us, the first to receive its seed from the Continent, and he dispersed it throughout Britain. The plants grown from it were esteemed for the shape of their flower stems, which curled "like caterpillars."

Scorpion grass was a natural candidate for the "doctrine of signatures," which attributed to a plant the property of healing ills of the part of the body that it resembled. Scorpion grass, with its curled raceme of flowers so like a scorpion's tail and its long history as an antidote for poisonous bites, was grown in herb gardens until the end of the seventeenth century, when John Ray, England's first great natural scientist, labeled the doctrine of signatures a superstitious folly.

Forget-Me-Nots in America

When the first settlers came to America scorpion grass was still an essential plant in the family herb garden. But as modern medicine developed, and home cures and herb gardens were abandoned, it escaped to the woodlands and waysides, where it proliferated as a wildling, and its ancient name and use were forgotten.

When the eighteenth and nineteenth centuries brought leisure and luxury to America and flowers were grown for pleasure, the *Myosotis scorpioides* bloomed in gardens under its old romantic name. No plant surpasses its sky-blue beauty and exuberant growth in the early spring border or rock garden. It blooms in six weeks from seed, it can be grown as an annual or biennial, and it makes an excellent ground cover.

This mouse-eared scorpion grass—for that is what its botanical name really means—is a member of the borage family, and is related to comfrey, anchusa, and heliotrope (which see). Its leaves are lance-shaped and hairy, its five blue petals merge in a yellow-centered tube, and its fruit is four minute curved seeds.

Distribution

There are many varieties of forget-me-not. The *M. sylvatica* is an annual, native to Eurasia, that likes the cool damp woodlands. The *M. verna* is a related American species that prefers dry ground and blooms in May and June. A brilliant dark-blue species is found in the Azores, but elsewhere is confined to greenhouses. It is the *M. scorpioides*, the Eurasian plant, that has become naturalized in America; it is familiar to our gardens and woodlands from Newfoundland to Georgia and westward to the Pacific coast.

Hybridists have developed strains with pink or white flowers. But it achieves its freest expression in Alaska, where the native variety, *Myosotis alpestris*, blooms in such abundance that Alaskan hillsides in summer are a brilliant blue.

Their beauty has made the forget-me-not the floral emblem of Alaska.

FORSYTHIA

The New England countryside is golden with forsythia. Every street, every garden, seems filled with its sunshine, announcing spring after the long cold winter—yet it was unknown in both America and Europe until the middle of the last century.

The story of forsythia is inextricably wound into the history of London's Chelsea Gardens and its far-seeing directors, whose interest in collecting and cultivating foreign plants made Chelsea one of the great botanical gardens of Europe. Phillip Miller, Chelsea's first distinguished curator, carried on a world-wide correspondence and collecting program for fifty years, and climaxed his career with the publication of the enduring *The Gardener's Dictionary*. He was a Scot, and he insisted on Scottish gardeners, who he said were a superior breed. His policy led, on his retirement, to the appointment of William Forsyth, another Scot and an even more flamboyant character than Miller. Forsyth specialized in shrubs and trees. His books on their care and diseases brought Chelsea international recognition and Forsyth himself an appointment as royal gardener at Kensington.

The cosmopolitan interests of Chelsea inaugurated a spacious age for English gardens. During the nineteenth century an endless stream of ornamental, economic, and medicinal plants flowed to England from Africa, Australia, and America—and then from China, which was at last reluctantly making itself accessible. When the first Opium War ended in 1842, the Chinese, by the terms of the Treaty of Nanking, opened five major ports to British ships and ceded Hong Kong to Britain. The prospect of importing China's floral treasures became irresistible.

Fortune and the Forsythia

The directors of the Royal Horticultural Society of London were eager to establish a resident English plant collector in China. They chose Robert Fortune, a young Scot, trained at Edinburgh's botanical gardens, who was in charge of the greenhouses at Chiswick. The experience particularly fitted him for testing the new device designed by Dr. Nathaniel Ward, the small terrariums in which plants could be sealed and shipped the 16,000 miles from China without watering—often a matter of life and death for a plant on sailing ships. (See Bleeding Heart.)

Fortune's assignment was neither easy nor safe. The Chinese resented all foreigners, and restricted them to the treaty ports. Fortune, however, was charming, able, and resourceful, and he accommodated his program to China's strictures. He cultivated the mandarins of the port cities, who graciously shared with him many plant treasures from their magnificent gardens. He became a customer of the great commercial nursery gardens, and paid generously for his purchases. And he visited Buddhist temples and cemeteries, where many rare plants grew, because horticulture was an essential aspect of Buddhism.

Even so, he found his every act supervised and suspect. When he made up his mind to explore the hinterland, he resorted to bribes, and to the extreme subterfuge of disguising himself as a Chinese. It was on one such adventure, in the spring of 1845, that he discovered forsythia. Dressed in native garb and pigtailed, with a crew of Chinese workmen, he explored the South China coast. The countryside was ablaze with a lovely bush whose every branch and twig were covered with bright-yellow flowers. It was growing wild; it was prolific and sturdy; and the Chinese called it "golden bell."

Fortune had his workmen dig some small specimens, which were planted in the compound of the East India Company until they could be shipped westward. Thus was the first forsythia (*Forsythia viridissima*) introduced to England.

He sailed into the Thames in 1846 with his terrariums packed with 215 healthy, new plants—the greatest living plant treasury ever brought from the East. Among them were several specimens of the Chinese golden bell, and many other plants now familiar to every American garden, including Chinese azaleas, camellias, weigelas, pompon chrysanthemums, bleeding hearts, and the Japanese anemone. The hardy, easily propagated golden bell was soon on sale at several nurseries, and in a few years was growing in English gardens.

Fortune was appointed curator of Chelsea Gardens, where he initiated a program of expansion and service in the tradition of Miller and Forsyth. In 1847 he also published a book on plant-hunting and horticulture in China, in which he described the cultivation of tea. And this led to new adventures and greater fame.

He returned to China a year later, primarily to collect tea plants for the East India Company, which was eager to introduce tea culture to India. (See Tea.) In the course of this mission he also collected many more plants unknown to the West. Among these was another forsythia, the golden rain tree, *Forsythia suspensa fortunei*, or weeping golden bell, whose graceful drooping branches, covered with yellow flowers, resemble a miniature shower of gold.

This forsythia is said to have been discovered in Japan in 1804, and shipped to the garden of one C. G. Breiter of Leipzig. It was noted there as an Oriental curiosity as late as 1817, but seems eventually to have been lost. Fortune's golden rain tree proved readily adaptable to England, and became quickly popular.

From Golden Bell to Forsythia

Fortune made other trips to the Far East, and in the twenty years he spent exploring and collecting in China, India, and Japan, other forsythias were sent to various English nurseries and became familiar garden shrubs.

But there was much confusion, and names for the several varieties were used interchangeably. The species needed an official botanical name and classification, and Fortune suggested "Forsythia" in honor of Chelsea's second curator. The name was a natural—there seemed an affinity between the aggressive, popular Scot and the hardy, glowing bush of early springtime—and by now "forsythia" has virtually replaced the old Chinese name.

Forsythia in the United States

Forsythia seems to have arrived in the United States in 1860, when *F. viridissima* was imported from England and planted in Massachusetts. Although it adapted well and propagated easily, the disruption of the Civil War delayed interest in its widespread adoption, as it did that of many other ornamental plants, and forsythia did not appear in nursery catalogues until the end of the century. *Forsythia intermedia*, which is the hybrid of *F. suspensa fortunei* and *F. viridissima* from which most modern cultivars are grown, was originally found as a chance seedling in the botanic garden at Göttingen, Germany, in 1878. Its larger, more numerous flowers brought it prompt recognition, and it was sent to the Arnold Arboretum in Massachusetts in 1889.

Like its cousin the lilac, forsythia has found a natural home in New England, and many spring safaris are made just to view the forsythia plantings on certain highways. Philadelphia gardens, too, became famous for their early introduction of forsythia. By 1910 it had spread to the Middle West, where it was often introduced on the campuses of land-grant colleges and agricultural schools.

In 1917 *F. ovata* was discovered in Korea, and was sent

directly to the Arnold Arboretum, where it first flowered in 1923. This shrub, only three to four feet high, is very hardy and needs far less pruning than other forsythias. In my garden, beside a carnelian cherry, it creates a magic all its own.

Botanical Bits

The forsythias belong to the Oleaceae or olive family. They are hardy, temperate-climate shrubs demanding the long rest of winter months, and do not thrive in Florida, the warm Gulf states, or the southwest. There are seven species, including a rare one found in Albania that has been much hybridized in recent years, producing some fine varieties. Most forsythias average eight to ten feet in height; all have opposite lance-shaped leaves about three inches long, and a profusion of golden, bell-shaped, stemless flowers, which cover the branches before the leaves appear.

Forsythia is now being investigated as a possible "green medicine." It has been found to contain rutin, an herbal extract that has proved beneficial in some kidney ailments and in capillary deficiencies.

Few shrubs provide more pleasure than this immigrant from the Orient. It brightens earliest spring days with a shower of gold, and enriches the summer with an abundance of rich green leaves. A well-pruned forsythia has grace and beauty, and a hedge of forsythia is quickly grown and guarantees an agreeable wall of privacy.

FOXGLOVE

In April 1775, William Withering, a country doctor in Shropshire, was working in his flower garden. He was an able botanist with an inquiring mind. He also had deep sympathy for his suffering patients, whom he was often unable to help because of the limited medical knowledge of the age.

Many people had as much confidence in their home-grown herbs as in a doctor's services. Take foxglove, for instance. How many times had he heard of that country woman who specialized in brewing up a mess of foxglove leaves and insisting it was "good for dropsy"? Foxglove was a powerful herb, and should not be handled carelessly. He must drive out there this very day and see what the story really was.

He found the woman intelligent, capable, and so convincing about the foxglove's efficacy that he decided to try her brew on one of his patients. The patient showed marked improvement and he decided to test it further. He continued experimenting with foxglove for the next ten years, and in 1785 presented a paper on the subject to a medical society. Many of the doctors present at the meeting subsequently tried foxglove on their own patients—also with success.

Four years later it was tested in Guy's Hospital in London, where it was used in certain tuberculosis cases, and during a scarlet-fever epidemic. Its amazing success in relieving a rapid pulse rate was duly noted, and many doctors began to keep case histories and to report their success in medical journals. Foxglove was hailed as a new advance in medicine.

Why, then, did it take another half century before its greatest value was discovered and it became the outstanding heart remedy of all time? Perhaps its very availability, and the fact that it was a folk medicine around which clung centuries of superstition, kept science from serious research into the elements in foxglove that have transformed it into a miracle drug.

Druid Magic

The native home of the foxglove is northern Europe and England, where it has grown wild for centuries. There the Celtic tribes wandered, worshiping the trees of their dark forests, and ruled over by Druids, who were at once the wise men and the medicine men of their people.

The Druids learned very early the power of foxglove to heal or harm. It is no coincidence that they claimed this tall, stately flower as their exclusive property, and used it to augment their own power. As if in uncanny co-operation, the foxglove stood out above all other flowers in royal colors of purple, gold, and white—each one the shape of a Druid's hat. With the foxglove the Druids performed almost magical feats of healing of various diseases. But the green leaves, used indiscriminately, can cause madness and even death, so they instilled in their people a fear of its poisonous properties.

Celt and Saxon, then, viewed the mysterious foxglove with awe—an attitude that endured through the ages and helped foxglove to maintain its important place among potent herbs long after the Druids became legend. Its fame as the most powerful of all plants for casting spells endured. So did its mixed reputation: it was known as an antidote for the aconite's poison, but many considered that it also belonged in the company of deadly nightshade.

How Foxglove Got Its Name

During the Middle Ages, when herb women assumed medical authority in isolated village and castle alike, the foxglove was still an important herb. Its prominence is attested to by the fact that it has sixty-two folk names in rural Europe, many of which—as, for instance, "witch's thimble" and "dead men's bells"—suggest its mysterious power.

The name by which we know it, innocent enough at first glance, is supposed to be a corruption of "folk's glove," meaning "fairy glove." The first mention of it in an Anglo-Saxon herbal, however, is as "foxes glofa" or glove, from the shape of its flowers.

Foxglove in the Herbals

As foxglove, it found its way into most of the north European herbals of the sixteenth and seventeenth centuries. The

German botanist Leonhard Fuchs gave foxglove its scientific name, *Digitalis purpurea*, or purple fingers, in his herbal *De Historia Stirpium Commentarii Insignes* of 1542.

The famed Flemish herbalist Rembert Dodoens, as well as the early English herbalists, including Parkinson, reported the medical use of digitalis, but none of them suggested it for heart conditions, and Gerard gave foxgloves short shrift, saying "they are of no use, neither have they any place amongst medicines."

As time went by the herbal repute of foxglove diminished, and by the late seventeenth century the English pharmacopoeia simply lists it as "a beneficial herb." This was the beginning of a long period of neglect for the foxglove, which became all but forgotten in the growing excitement over the endless new plants pouring into Europe from the New World and the East.

Foxglove in America

During the colonial period the foxglove is absent from all ship lists of herbs essential for import, and there are no records that it grew in the gardens of early settlers. It is not until 1760 that we find any mention of the plant in America. In that year John Bartram reports that seeds of yellow digitalis have arrived in a shipment from England, and have been planted.

After the Revolutionary War, one of the first reuniting links between England and America was the exchange of medicinal plants, and foxglove played an active part. This is recorded in a letter to Ezra Stiles, president of Yale College, from his old friend Dr. Hall Jackson, of Portsmouth, New Hampshire, in 1787. Jackson writes that he has just received a quantity of foxglove seed from the famed Dr. Withering, who asks him to share the seed with interested men in other states who would cultivate it and spread the knowledge of this beneficent plant and its medicinal use.

There is, however, no doubt that both English and German colonists brought foxglove to America very early. They scattered its seeds among the stumps of their cleared land on the edge of the forests, where it throve mightily, and they gathered its woolly leaves to use as poultices for persistent sores and swellings, as their forebears had done in Europe. Other settlers let it grow because it comforted their homesick hearts and could be relied on to come up from seed year after year, to brighten their gardens without demand on their hard-pressed labor.

Foxglove soon escaped from such homesteads on the fringe of the continent and began its trek westward. So rapid was its progress that by 1809, when the English botanist John Bradbury was sent to America to collect new floral treasures in territory opened by the Louisiana Purchase, he found the land covered with an infinite variety of wildflowers, and reported in amazement that where pioneers had passed, an English flower, the foxglove, was growing among them.

The foxglove is designed for traveling. There are 80,000 seeds to an ounce, and the wind tears them from their papery capsules and scatters them lavishly each fall. They are rich in oil, starch, protein, and sugar, which assures a high rate of germination.

Since foxglove is essentially a biennial, it seizes squatter's rights to the land by forming a rosette of woolly leaves the first year, and takes full possession early the next spring by putting up a commanding flower stalk from three to five feet tall. Large thimble-shaped flowers grow along this stem above the leaves—and it is these second-season leaves that contain the powerful chemical that is called, like the plant itself, "digitalis."

The flowers, which may appear from early spring through June, are pollinated by wild bees. Each flower wears bright spots inside its throat to guide the bee to its nectar, and it fits the bee so neatly that he brushes off the pollen from another flower as he feasts.

There are two main strains of foxglove, which can easily be distinguished by leaf coloration. The foliage of the purple and pink foxgloves has dark veins; that of the white and yellow, light-green ones. The common old magenta is the hardiest for naturalizing, and has most often escaped to the roadside. Many consider it the most beautiful, as well.

Modern Foxglove

In 1810, after a long experimental period following Withering's rediscovery of digitalis as a powerful natural drug, the medical profession cautiously announced that digitalis was "a new medicine, used experimentally and not yet understood." The foxglove then entered upon an era of recognition and widening use.

At last, in 1871, a French scientist named Nativelle isolated the active chemical in foxglove leaves and called it "digitalin." From continued experimentation, doctors learned why the substance relieved dropsy, which was caused by weak heart muscles. When the heart was unable to pump fluid through the veins, it accumulated, and the body swelled painfully—a condition digitalis relieved by helping the damaged heart to work more efficiently. Except for the lily-of-the-valley, there is no comparable mender of hearts.

As a result of the universal use of digitalis, several new strains of foxglove that are richer in glycosides, its natural healing property, have been developed. Their cultivation for medical use is rendered expensive by today's higher labor costs, because much of the planting and seed-collecting must be done by hand. Our principal commercial beds are in New England and Oregon.

Foxglove plays another important role in the modern world. Back in the sixteenth century, Dodoens observed that foxglove was found in valleys where there was much mining of iron and coal. Today Russia prospects for new iron and coal fields from helicopters, searching for masses of foxglove, which indicate the presence of such minerals in the ground beneath.

New decorative strains of foxglove have also been developed. Some are perennial and others are available in wider color variations with larger flowers, on taller stalks, which may reach six feet. Majestic and beautiful, they grow in gardens around the world, still creating an aura of mystery, still commanding homage and respect.

FUCHSIA

Centuries of intrigue, exploration, and romance lay behind the fuchsia's late debut in European gardens, where it reigned as the most popular flower of the Victorian era.

In 1556, Spain declared a ban throughout Spanish America on exploration by all plant hunters. The restriction, which included total secrecy about discoveries by Spain's own naturalists, lasted, with only a few exceptions, for nearly 300 years. Moreover, it provoked Englishmen like Raleigh, Drake, and Hawkins into raiding not only Spain's treasure ships but also her vulnerable island ports in tropical America; so they carried home to England, along with their other booty, plant treasures such as the potato, the pineapple, and the wallflower, as inducements to the monarch to finance further raids.

In 1654, a year after the establishment of the Protectorate, Oliver Cromwell authorized a naval expedition to the West Indies, under the leadership of Admiral William Penn, whose son was to become the founder of Pennsylvania. Penn captured Jamaica from the Spanish, thereby establishing a permanent foothold for Britain in the Antilles, and opening the way for the migration of Caribbean plants to English gardens.

Some forty years later Louis XIV hit upon a way for France, in her turn, to acquire a base in the Caribbean. Charles II of Spain, crippled mentally and physically, was dying childless, and the French king's grandson was heir to the Spanish throne. Louis offered his aid to Spain against English aggression in the West Indies, with the proviso that Spain officially cede to France the western part of the island of Santo Domingo, or Hispaniola, which now constitutes Haiti.

As soon as the agreement was signed, Louis sent off Father Charles Plumier to explore the island and collect its plants. A few months later he also sent out young, able Michel Bégon, ostensibly to be governor, but, secretly, because he, too, was a botanist, who had collected new plants from every part of the world where his ships had landed. (See Begonia.) It was in Santo Domingo that Father Plumier discovered fuchsias, as he and Bégon explored the island in search of plant treasures.

Plumier described this hitherto unknown flower as possessing rare color and form, with a pendant blossom of red and blue with four calyxes, four petals, and four stamens. He named it *Fuchsia triphylla coccinea*, in honor of the German botanist Leonhard Fuchs, whose new scientific approach to the study of plants had greatly influenced him.

There is no record that Plumier's fuchsia was cultivated in Europe. If he sent or took back living plants, they were lost; but he did include a detailed drawing and description of it in his famed volume *Nova Plantarum Americanarum Genera* (New Families of American Plants), published in 1693. This book contained descriptions of fifty other new West Indies flowers that Plumier had named for famous Europeans, including the begonia.

The Spanish government loudly protested Plumier's revelations, which aroused the envy and ambition of other countries and inspired further botanical expeditions to the New World. Spain saw to it that the activities of foreign plant hunters were restricted and supervised whenever they attempted to go beyond the limited possessions of France and England.

But now France found another way to wedge open the door to Spanish America. Father Louis Feuillée, like Plumier a member of the Order of Minims and an outstanding scientist, was dispatched with the French fleet to the west coast of South America, officially to make astronomical and climatic observations for the French navy, but in fact to map the Pacific harbors and record in word and picture the natural resources of the Andes.

Feuillée managed to linger in the area for some years, and later published, 1714–25, his superb *Journal des observations*, which rivaled Plumier's in the number of botanical discoveries it reported. It included, among other things, fifty plates of medicinal plants. It also described a new fuchsia, which differed completely from Plumier's. But Feuillée was not a collector, and only the illustration and description of his fuchsia reached Europe. His book, however, further whetted the appetite of the plant hunters, who now accompanied most military and naval expeditions.

Fuchsias in England

Another half century passed without further word of fuchsias. None were known to be cultivated in Europe, and they had become something of a botanical myth. And then a fuchsia was discovered in England.

Although the story of its journey there varies, the following version seems the most logical. In 1788 a naval captain by the name of Firth, returning from the Orient, had been forced to harbor in Chile for repairs to his ship before undertaking the voyage around Cape Horn and home across the Atlantic. As he wandered along the shore he came upon a remarkable plant whose like he had never seen before. Its bright-red pendant flowers and long lacy stamens were so appealing that he dug up a small specimen, took it to his ship, and cared for it in his cabin. When he reached London he presented it to his mother, who put the plant from foreign parts in the window of her house in Wapping on the Thames. There it flourished, and finally it bloomed.

One day James Lee, of a famous vineyard nursery in nearby Hammersmith, observed the exotic plant, and, having learned its story from Mrs. Firth, persuaded her to sell it to him for the fabulous sum of eighty golden guineas.

James Lee knew a treasure when he saw one. He sent the plant to Kew Gardens, where the director, William Aiton, a distinguished English botanist, pronounced it to be a new species of *F. triphylla coccinea*. Lee had found that it grew happily in his greenhouse, and propagated easily, and he soon sold 300 cuttings at a guinea each. *F. coccinea* made his nursery famous, but it could not adapt to the English climate as a garden flower. When the *F. macrostemma* was introduced from Chile in 1838, and proved to be hardy and delightfully variable, it soon overshadowed *F. coccinea*.

During the nineteenth century the increasing ease of communication and travel eroded the isolation in which Spain had kept her American possessions. Though she still harassed plant explorers with delays, deceits, and impossible conditions, she gradually lost control of the nations of Central and South America—and fuchsias poured out on the world.

In 1840 *F. splendens* was found in Guatemala, and plants and seeds were imported to England. It proved easy to propagate, and was soon growing in gardens throughout the British Isles. In 1898 *F. fulgens*, a broad-leafed variety discovered in Morelia, Mexico, by Theodore Hartweg, brought the fuchsia period to a climax.

Few flowers have enjoyed such prolonged popularity. The fuchsia was the rage of England's conservatories and the pride of its cottage window sills, and its jewellike flowers won the familiar names "lady's-eardrops" and "fairy bells." Up to the outbreak of World War I the flower markets of England were always full of fuchsias in great variety; but as a part of the nation's defense efforts the gardens where they had grown were dug up and converted to the cultivation of food crops.

The fuchsia's popularity had been slipping away in any case. At the beginning of the twentieth century 1,500 varieties were known in England. Today that total has dwindled to a few hundred, and only a few dozen varieties are actually grown. Hybridization, whose techniques were developed in this period, had overtaken the fuchsia, crossing its dissimilar strains and making its variability commonplace.

Fuchsias in Today's World

The study of plants expanded in the second half of the nineteenth century, and became a specialized career, recognized by colleges as having commercial importance: botany was no longer just a recreation for amateurs. The change intensified the search for plants of economic value, and for exotics for commercial greenhouses.

Among the latter-day plant hunters, Benedict Roezl, trained in Belgium's Van Houtte nurseries, has what may be the most impressive record of all time. He tramped countless hundreds of miles, from Mexico to southern Peru, and from Cuba to California and Vancouver. He collected thousands of rare plants—lilies, orchids, calceolarias, and several new fuchsias—which he shipped to European nurseries in wholesale quantities. Viewing this avid plunder of the wilds for profit, one almost wishes the restrictions of the Council of the Indies still held.

Two other twentieth-century plant hunters who contributed to the fuchsia's history are Clarence Elliot and H. F. Comber, who discovered the lovely white fuchsia, *F. magellanica*, on a trip to South America in the late 1920's.

Fuchsias in the United States

Fuchsias remained a greenhouse plant in our nation even at the height of their popularity in English gardens, and received little attention here until after the Civil War. Then, in 1873, a

New York nurseryman named Thomas Hogg received from Haiti some seeds that produced a fuchsia quite unlike any previously known in America. Eventually he sent seeds from it to Messrs. Henderson & Sons, who were fuchsia specialists; they, too, were unable to identify the variety, and sent the seed on to Kew Gardens. There plants grown from it were recognized at last as the original *Fuchsia triphylla coccinea*, which Plumier had found in Santo Domingo two centuries before and which had not been seen since!

All of this helped to bring fuchsias to the attention of the American public and the commercial florists. They became popular as house plants, and in pots or hanging baskets soon bloomed on thousands of winter window sills and on millions of summer porches and patios. But fuchsias do not grow in gardens above the frost line, and they have never reached the popularity here that they knew in Victorian England.

Fuchsias, however, grow in wild abandon in California. On the coast north of San Francisco and into southern Oregon, the masses of fuchsias are breath-taking. These are not modern hybrids like our house plants, but relics of an earlier day when the region was part of Spanish America and fuchsias were planted as hedges around the haciendas. Today, free in the wild, they grow six to eight feet tall, their brilliant bell-like flowers attracting ruby-throated hummingbirds by the hundreds.

Fuchsias in Hawaii

Fuchsias were carried to Hawaii a century ago and found a perfect home there. They soon escaped from gardens and parks, and by now grow wild throughout the islands.

An unforgettable memory of a Hawaiian holiday is of stepping from a misty fern forest into a sunny pathway arched with vines and wreathed with fuchsias in full bloom. With what exquisite artistry nature had hung these tiny lanterns, their brilliant color catching the sun as it shone through the mist of the green jungle. I was tempted to believe this fleeting moment of beauty had been designed just for me.

Botanical Bits

The fuchsias belong to the Onagraceae family, a humbler member of which is that familiar weed of roadsides and waste-lands, the evening primrose. Most fuchsias come from tropical America, although a few minor species have been found in New Zealand and are thought to be native there. Some are climbers, some are shrubs and even small trees. All of them have exquisite axillary flowers that are solitary and pendulous, though they sometimes bloom in terminal racemes. The flowers have four cleft calyxes and four petals. The fruit is a berry of four cells, and in a few species is considered edible.

Hybrid fuchsias may produce either single or double blooms, which come in many shades of the basic colors of red, white, and blue. The flowers are always nodding and showy and the leaves opposite, oval, finely toothed, and graceful. Few plants are endowed with such individual beauty, and few require less care. They well deserve a return to popularity, and their devotees are doing something about it. Today's American and English fuchsia societies are working to pro-mote fuchsia-growing and to restore many old varieties to popularity among amateur and professional gardeners—a delightful ambition!

GERANIUM

One day in 1632, after a stormy voyage westward across the Indian Ocean from Surat, and a difficult passage around the Cape of Good Hope, an English sailing vessel came to anchor in one of the numerous small rivers that intersect the coast of southern Africa. While the ship was being repaired and water taken aboard, the captain went ashore to investigate this area, which was claimed by the Portuguese, who had discovered it, and by the Dutch, who were attempting to settle it.

He found forests of yellowwood, ironwood, and palms along the coast; beyond rose plateaus covered with grass and bush, called by Dutch settlers the "veld," which seemed rich in unknown trees and shrubs and flowers. He took back to the ship a plant that had especially caught his fancy—a woody herb with a ball of bright flowers and beautiful scented leaves. It filled his cabin with a fresh fragrance for the rest of the long voyage home.

Thus did the first South African geranium come to England. We can safely assume it caused a great stir in London, where all new foreign plants were given a royal welcome. It throve in the damp, mild English climate, and from that time on traders, missionaries, and sea captains, both Dutch and English, brought back from South Africa a never-ending succession of geraniums of every size, color, and variety.

That original geranium was lost, and the details of its introduction grew vague, but it initiated a search that has lasted for three centuries.

How the Geranium Got Its Name

One of the first questions to be settled when a new plant arrived in Europe was what to name it. The name "geranium" comes from the Greek word for "crane." It had originally been used in the first century by Pliny the Elder, who had given the name to a class of plants native to Europe because their long pointed seed cases suggested a crane's bill. Pliny's crane's-bill is the true geranium. It is endemic to the Temperate Zones of the Northern Hemisphere, and grows variously as an annual, a biennial, or a perennial in the fields and woodlands of Europe and North America. The blossom of this wild geranium, which consists of a loose head of purple-pink florets with five regular petals, is far less showy than its South African relative, but surely one of the airiest and most graceful of wildflowers; and in its long history it has earned a number of picturesque folk names: "herb Robert," "Robin redshanks," "meadow crane's-bill"—and wild geranium. Several cultivated dwarf varieties are now available, and make charming rock-garden plants.

When the South African species reached Europe, botanists noted its similarity to the wild native geranium—and bestowed that name on the showy newcomer. The stir caused by these African geraniums injured the pride of the Dutch, who accused the English of invading the African veld, to which they laid claim, and stealing its wildflowers. The geranium became, indeed, an international issue that smoldered for nearly a century.

Holland began importing several species in 1690, and clinched her claim to the African geranium with the publication, 1697–1701, of an herbal, *Horti Medici*, by botanist brothers Jan and Caspar Commelin. This contained the first official description of the geranium and the first picture of it, a water color by Johan Moninckx. One of the loveliest herbals to come from the early Dutch presses, it may still be seen at the Amsterdam Botanical Garden.

But Holland's claim to the geranium was short-lived. In 1710 a ship arrived in London from Cape Town with many new species, which removed them from the rare-plant class and provided the basis for their unprecedented popularity as garden flowers. For this reason, although they had already been well established among botanists, 1710 is often given as the year in which geraniums were first introduced to England.

By 1732 one English gardener boasted that he possessed six different African geraniums. Twenty years later Linnaeus was describing twenty-five known varieties, and by 1789 L'Heritier published his claim to a collection of forty-four.

L'Heritier, a true scientist, decided that the growing number of varieties should be classified and their relationships established. This he did by recognizing Pliny's European crane's-bill as the true geranium and by placing the South African species in a new genus, which he called *Pelargonium*. The name derives from the Greek *pelargos*, for "stork"— thereby stressing the pelargonium's close relation to the European crane's-bill, especially in the shape of the seed pods.

L'Heritier subsequently found that there were a number of geraniums that did not fit either of these classifications and established a third genus, which he named *Erodium*, or "heron's-bill." There were fewer than half a dozen erodiums known in L'Heritier's day, but by now nearly fifty varieties have been found in Asia, Australia, and Africa. It is an enchanting genus, and provides many miniature plants for modern gardens; a few species, salt-air lovers, are found among the sand dunes of New England. Nearly all the erodiums closely resemble wild geraniums, except for the terminal flowers, which grow in groups of from three to seven.

Botanically, L'Heritier's classification of the three branches of the family Geraniaceae is invaluable, but to the average gardener a geranium is a geranium is a geranium. Adapted to growing both indoors and out, all over the world. it has been a joy to generations of such gardeners.

Geraniums in America

The early arrival of geraniums in America is recorded in a letter dated 1760 to John Bartram of Philadelphia from his London associate Peter Collinson, which contains a receipt for a shipment of seeds from England—among them geranium seeds. Bartram, who was one of the first links in the exchange of plants between Europe and America, soon became a geranium enthusiast, and was quick to spread word about this new favorite. A craze comparable to the one in Europe seized the colonies, and for the next fifty years geraniums led all other garden flowers in popularity.

There were, however, few American nurseries in colonial days. The owners of large estates and gardens imported everything from Europe, and thousands of colonists carried on a correspondence with horticulturists, gardeners, and naturalists in the homeland. A few geranium seeds tucked into a letter from Europe might be producing blooms in the New World five to eight months later. And once the geranium was established over here, it could be rapidly propagated by cuttings, air layering, and grafting, a factor that won it instant appreciation. There seemed, as well, an endless variety to its colorful flowers and fragrant leaves: it was a deeply satisfying plant, which provided the maximum of joy for the minimum of care.

For such reasons it was peculiarly fitted to American pioneering. Hundreds of covered wagons carried geraniums across the country, and, like the people, they quickly adapted to new climates and soils. In California they now run riot over the hills, and are used as ground cover on steep banks and as hedges for public parks and private gardens. There, too, the standard tree geranium and the ivy-leafed topiary geranium are grown in tubs in walled gardens and patios, scenting the air and producing an abundance of blooms.

Geraniums were so intertwined with America's rural life during the last century that they became synonymous with home and hospitality. Exchanging geranium slips—between individuals, in garden clubs, or at church bazaars—became a country custom, like drinking cider. "Garden Open Today" was an invitation for a carriage drive and an afternoon spent viewing a geranium collection. And pots of geraniums shone in the flower stalls at country fairs, rivaling the exhibitions of prize vegetables and livestock. There they stood—pink and red zonal geraniums and Lady Washington regals, many of which had been in the United States as early as 1840; the scented-leaf ones, and the miniatures; the singles and the doubles in reds, and pinks, and white.

Geraniums were utilized for cooking, too. In the days when spices were scarce and expensive, they provided cinnamon, nutmeg, and lemon-scented leaves. Quince preserves and apple jelly gained piquancy from a rose-geranium leaf in the bottom of each glass, and even roasts were seasoned with a touch of spice from the geranium pot on the kitchen window sill.

The Crest of the Geranium Craze

The geranium craze reached its height in England in the early 1800's with the publication of Robert Sweet's *Geraniaceae*,

1820–1830, and the creation of Lady Grenville's geranium garden.

Robert Sweet's work, a labor of love that occupied a lifetime, contains 500 exquisite colored plates of geraniums, and is splendidly informative. It is still available in many state libraries, and at large universities and botanical gardens. Once you have found it, its serene beauty and its historical and botanical fascination can hold you captive for hours.

As for Lady Grenville, she landscaped her garden with a prodigious number and variety of geraniums, using them with striking ingenuity. The immediate result was a new craze for the lavish use of geraniums in gardens, large and small, throughout England and America.

But geraniums were propagated with such ease and success that even rare species became common, and in time their popularity began to fade.

Geraniums as Perfume

It was just at this point that French perfume manufacturers discovered that geraniums constituted an inexpensive source of scented oil. For centuries the rose had provided a variety of fragrances that ranged from spicy to fruity. Now, suddenly, perfumers discovered that geraniums could provide even greater variety at a far smaller cost. As a base for perfume, the geranium has unlimited possibilities, because it can be hybridized to create new scents, and it is easily and quickly grown.

Before 1850, Turkey and France were growing acres of scented geraniums, which placed them among the plants of high commercial value to man. It also meant that perfumes, once a luxury for the rich, could be mass produced and so become available to all.

Geraniums of the Future

The commercial use of geraniums has led to continual experiment and development, and to mechanized cultivation, harvesting, and sorting. Geranium seed is small, feathery, and light, designed to be carried by the wind; but today it is gathered by mechanical vacuum pickers, which take up seed and waste, half and half. The seeds are sifted through great automatic vibrating screens, and passed on to robot packers that count them and box them for shipment.

The great American farms where geraniums are grown and bred for wholesale markets, for seed or for experimental purposes, are located mostly in southern California, in arid country near the ocean, an environment close to that of the Cape region where they were discovered. Their commercial cultivation has created new interest in growing them for pleasure. National and international geranium societies list thousands of members and publish monthly fliers and magazines.

The geranium now seems peculiarly an American flower. Its familiar fragrance and beauty spell home to thousands of Americans living abroad, especially in Europe. And most of the new varieties are developed in the United States; Cornell University, in particular, has altered and improved a number of old favorites, which appear under new names.

But many nations feel possessive about this South African wildling, and because of world-wide interest in it hybridists and nurserymen predict that a geranium of totally new dimensions may become the symbol of tomorrow's One World.

GLADIOLUS

Gladioluses are said to be the lilies of the field that Jesus referred to in the Sermon on the Mount. They were once familiar weeds in the grain fields of the Holy Land, and grew wild in wastelands along the Mediterranean coast of Africa.

Centuries later they made an entrance into England in dramatic circumstances.

In 1620 Charles I ordered a naval expedition to the African coast to rout the Barbary pirates who were harassing British ships. John Tradescant, Sr., royal gardener to Charles I, persuaded the king that this would provide an excellent opportunity for gathering African plants, and received permission to accompany the fleet. Between battles he explored the Barbary Coast, and collected many new flowers, including the Byzantine gladiolus, which, he noted, grew in great quantities near the shore.

A few specimens had been introduced as early as 1578, but the "glad" or "corn flag" was not widely known in England, and gardeners gave the new arrival a cordial reception. John Parkinson notes that the Byzantine gladiolus was thriving in his garden in 1629, but the history of the genus in England remains vague for some years thereafter.

In 1652 the Dutch East India Company established the first permanent colony on the Cape of Good Hope, a strategic location for servicing their ships on the long journey to the Far East. Among the first colonists at the Cape were Flemish Huguenots, who were famous as the best horticulturists of Europe. As these pioneers spread out from the coast of the Cape Colony to the karoo and veld—dry plateaus and grasslands—they found that the whole area was rich with a wide diversity of flowers, among them the Cape gladiolus. The Flemings knew how highly European gardeners would value these plants, and soon cargoes of them—gladioluses, calla lilies, aloes, euphorbias—were carried to England and Holland by every ship from the Cape Colony.

European plant hunters began exploring the Cape. New species of gladiolus were among the greatest treasures they found and introduced to nurseries in England and on the Continent. The demand for these plants grew steadily through the next seventy-five years.

They were familiarly called "sword lilies" from the shape of their leaves. When Linnaeus classified them in 1770 he gave them the official name *Gladiolus,* the diminutive of the Latin word for "sword," and placed them in the order of Iridaceae, the iris family.

In France, Victor Lemoine (1823–1911), greatest of the French plant breeders, who worked in Nancy, began the hybridization of gladioluses, and developed the lovely "pansy" strain, which he named for the dark velvety markings that were reminiscent of pansy faces. In 1841, in Belgium, the well-known nurseryman L. B. Van Houtte originated the *gandavensis* variety, which, being both beautiful and easy to grow, is one that has greatly helped to make the gladiolus a favorite garden flower.

In England the *Gladiolus aethiopian* and the fragrant *G. tristis* were reigning at Kew Gardens and Chelsea Gardens when Collingwood Ingram, a latter-day collector, returned from the Transvaal with a hundred new varieties, many of them from the farm of General Jan Smuts, who was an inveterate naturalist. Among the most important of Ingram's finds was the *G. lilaceus,* whose blossom changes color from brown in the daylight to blue at night, and back again to brown the next morning. This oddity, although first found in 1770, was little known before 1920, since when it has become widely popular as the "brown Africander."

Another late discovery was the primula gladiolus, which was introduced in 1908 and caused a great stir. The abundance and the brilliant yellow of its blooms gave fresh impetus to the hybridists, and innumerable new varieties and color combinations were created and marketed.

Gladiolus in America

In spite of the furor in Europe, gladioluses were ignored in earlier American garden books and catalogues. The Philadelphia seedsman Bernard McMahon is the first to mention them: in some notes on growing bulbs in American green-houses, dated 1806, he urges the introduction of the Cape gladioluses. He was, incidentally, a friend of President Jefferson, and planned with him the plant-collecting aspect of the Lewis and Clark expedition. New plants discovered on the expedition were sent to McMahon's nursery, and many of them were later introduced to American gardens. His interest in gladioluses endured, as we know from a letter he wrote to Jefferson in 1812, to say he was doing himself "the pleasure of sending you by mail . . . 12 roots of gladiolus commensus."

The first catalogue to offer gladioluses to American gardeners came from the New York nursery of Grant Thornburn, a canny Scots immigrant who had started a career by investing his last sixpence in three pots of geraniums, which he peddled at a profit. Now highly successful, he offered his gladioluses as "South African Sword Lilies," at twelve cents a bulb. Soon thereafter the Landreth Nurseries, located in Philadelphia, offered two varieties of sword lilies, and the great Prince Nursery of Flushing, Long Island, advertised "Corn Flags or Sword Lilies" in four colors. But American gardeners in general appear to have remained indifferent to the gladiolus.

In 1836 a new attempt, from quite another source, was made to acquaint America with this dramatic flower. Baron C. F. H. von Ludwig, who owned a fabulous garden in Cape Town, sent a collection of various species of gladiolus to the Massachusetts Horticultural Society—perhaps at the instigation of his English gardener, James Bowie, a man of wide renown and the last of the great plant hunters to collect specimens of Cape plants. The Horticultural Society planted the gladioluses at their experiment station, and later organized for its members an exhibit and lecture on gladiolus culture.

One convert won by Ludwig's effort was Francis Parkman, who besides being a historian was an ardent horticulturist. He became a gladiolus enthusiast, and put on a show for the floral committee of the Horticultural Society at which he exhibited the many species and varieties that were available.

But the Civil War had then started, and apart from gestures like Parkman's little interest was shown in gladioluses. The *American Gardener's Magazine*, after a survey of Boston gardens at the time, reported that only two were growing them.

Burbank's Hybrid Gladiolus

Around 1878 Luther Burbank began breeding gladioluses at his experimental farm in California. The climate was similar to that of the Cape, and the Burbank gladioluses flourished. Twelve years of experiments followed, and in 1890 he put his first hybrid gladioluses on the market, offering an assortment of ten varieties for ten dollars. But he asked two dollars a corm for a few especially choice ones, a goodly sum at the time.

It was also in the nineties that H. H. Graf, of Ontario, Canada, after years of rigid selection, developed some superb hybrid gladioluses that became widely known. The success of the Burbank and Graf hybrids was followed by the development of the ruffled and fluted strains at the Kunderd Nursery in Goshen, Indiana. Other nurseries, on Long Island and in the south, next caught the gladiolus fever, and began hybridizing and producing bulbs on a commercial scale. Then Dr. Forman T. McLean undertook a thirteen-year program at the New York Botanical Garden, in the course of which he created the soft-colored hybrids which, unlike most earlier ones, were scented, and which he named "sweet glads."

Botanical Bits

More than 200 species of gladiolus have been found on Africa's Mediterranean coast and in South Africa combined, but the greatest concentration is in the Cape Colony.

The gladiolus is a cormous plant, whose corms develop best when the seed spike is removed after flowering. Cormels are produced on underground stems running from the base of the corm—in general, most abundantly from corms that have not yet put up a flower stem, although this is not true of all varieties. A single corm may produce 200 cormels, each of which is enclosed in a hard, fibrous shell.

Some South African tribes dry the cormels and pound them into a flour from which they make bread. And there are natives of the Mediterranean area who value them as medication for various ills.

The gladiolus possesses the characteristic stiff, narrow leaves, rising from the base of the plant, of the iris family. The flowers, on tall, spiked stems, begin blooming at the bottom and open upward, a few at a time. They vary in size from an average two inches; they are irregular, colorful, and of great appeal. The flower stem has the capacity to take up water and hold it, enabling the flowers to bloom to the very tip of the spike even after they are cut, which increases their value for florists.

Although amateur gardeners generally rely on corms, gladioluses can also be grown from their winged seeds. The fact that seed-grown plants do not bloom true to form is an asset to the hybridizers and has helped them to produce the splendid modern gladioluses that result from extreme crossings.

When the first gladioluses arrived in the United States, American seedsmen were at last independent of European nurseries, and were able to develop hybrids suited to our wide range of climates. They created thousands of varieties in a great number of colors and types, many far superior to any abroad, all of them different from European and Australian species and none like the African wildlings of a century ago.

The tables have been turned. Before World War I southern Europe supplied the world with gladiolus corms.

Now the United States produces most of them, not only for American gardens but also for seed houses and gardens abroad.

A Matter of Pronunciation

For two centuries the sword lily of Africa has been called "glăd ĭ ō lŭs" by most gardeners, but modern experts now prefer "glá dȳe ō lŭs." However he pronounces it, glad is the gardener who grows gladioluses—not just for their own sake, but because he can be sure of a rich dividend of humming-birds, who come seeking their nectar.

GRAPE HYACINTH

The grape hyacinth is used as an ingredient in Mediterranean cooking, and great quantities of its bulbs are exported to America each year as a delicacy for Italian-Americans. This amazing statement is found in a famous Italian plant book, *The Complete Book of Fruits and Vegetables*, by Francesco Bianchini, Francesco Corbetta, and P. Marilena, recently republished in the United States, which recommends that the bulbs of varieties *Muscari comosa* and *M. atlanticum* be boiled with a little vinegar to reduce their bitter flavor and then made into pickles.

It is hard to think of the delicate blue grape hyacinth, one of the earliest heralds of spring in the garden, being devoured as a pickle. But it has grown in European fields for centuries, and apparently must pay tribute by feeding man.

The grape hyacinth is native to southern Europe, North Africa, and western Asia—the Mediterranean littoral—but had spread through the rest of Europe to England before recorded history. An early Anglo-Saxon herbal implies its specific medicinal use in the name *Bulbus vomitorium*, but later herbals recommend it as a diuretic and an appetizer.

Many more varieties were grown in Elizabethan gardens than are known today, and both Gerard and Parkinson describe them in some detail. Gerard calls the species *Muscari comosa*, "faire-haired" and remarks on its "musky perfume." Parkinson aptly observes that the *Muscari botryoides* smells like hot new-made starch, and even today the folk name "starch hyacinth" is commonly used for it.

Grape Hyacinths in America

When the first Europeans settled along the Atlantic coast, grape hyacinths were at their height of popularity, growing in both great and small gardens. Because bulbs could be transported easily on the small crowded ships, and because grape hyacinths were evocative of the homes the settlers had left behind, they found a spot in New England, New Netherland, Pennsylvania, and Virginia alike. But the "heavenly blue" grape hyacinth of our modern gardens owes much of its popularity to Peter Barr (1825–1909), an English nurseryman who brought home with him a few large bulbs of *M. botryoides* he had found in Trebizond, on the Black Sea. Under cultivation they spread rapidly, and developed colonies of robust plants adapted to sunny fields as well as to gardens. Their vigor and bright color soon made them popular in the Victorian era, and they even inspired a description by Ruskin as "a cluster of grapes and a hive of honey—distilled and pressed together in one small boss of beaded blue."

In the last two centuries the grape hyacinth has made itself at home across America. Pioneering in various climates and soils, it has escaped from cultivation to lawns, fields, and roadsides; but our nurseries continue to compete against the runaway with new cultivated varieties.

Biography of a Grape Hyacinth

The grape hyacinth belongs to the lily family. There are more than forty varieties, of which the best known remains the "heavenly blue" *Muscari botryoides*. It takes its botanical name from the Greek words *moschos*, or "musk," and *botryoides*, which is Greek for "a bunch of grapes."

The grape hyacinth is one of the most enduring and endearing of spring bulb flowers. In earliest spring it sends up slender grasslike leaves, with no center groove, from which spring the flower stems bearing racemes of blue urn-shaped flowers. The top florets, usually showy and sterile, attract the insects that carry the pollen to and from fertile flowers.

The whole plant seldom exceeds six inches in height, and resembles a miniature Oriental hyacinth. Although there are pink and white varieties, and double hybrids, the typical color is the bright blue of spring skies. Once planted in quantities, they ensure masses of color in the early garden for years to come. Perhaps this is the secret of their enduring popularity— a joyous surprise each spring for winter-weary humanity.

"How carefull·diligent
I have been,
These coloures
to express:
In painfull paintings
of the same
Good reader
use no lesse."

Salomon de Roy
for
Crispin de Passe
Utrecht
1615

HELIOTROPE

After a year of ambitious planning and organization, and of prolonged travel, a committee of French scientists arrived in Quito, Peru, in 1736. They had come ostensibly to measure the degree of latitude at the equator, which had never been definitely established. Spain had consented to this project with great reluctance, being convinced that the true purpose of the French was to plunder Spanish America's wildlife. She had insisted that two Spanish commissioners supervise the expedition and that its activity be confined to geophysics.

At Portobelo, the old fever-ridden gold port of Panama, the committee was delayed many months, under close observation by the Spanish. Just as the latter had foreseen,

Joseph de Jussieu, a brilliant physician and botanist, and brother of Antoine de Jussieu, of the Jardin des Plantes, who was attached to the expedition, could not resist exploring, and collecting seeds, bulbs, plants, and a herbarium. But Jussieu partly redeemed himself in Spanish eyes by his dedicated service when an epidemic of tropical fever broke out; and the Frenchmen were at last permitted to leave.

Three years passed after they had established a camp in the Andes, near Quito, and begun their geophysical studies. Jussieu, who could no longer resist the opportunity to botanize, went off to examine the plants from whose bark quinine was derived: the French were familiar with the medicine, but knew nothing about the family of shrubs and trees called *Cinchona*. After another three years, the camp broke up, but Jussieu decided to remain in the Andes and pursue his solitary botanical investigations. Months later, as he wandered alone through the valleys of the Cordilleras, he was arrested by a poignant fragrance that he traced to a weedy field flower somewhat resembling the European garden heliotrope. This pale, trailing wild Andean herb was less showy, but its fragrance "intoxicated with delight." He called his discovery "Peruvian heliotrope."

His investigations continued for another year. He slowly made his way southeast toward Buenos Aires and a ship for home, burdened with his rich collection, his herbarium, his notes, and his sketches—the fruit of years of lonely work. And then, as he neared La Paz, the native servant who had traveled with him abruptly disappeared into the Bolivian hinterland, taking everything with him. One wonders what use he made of these purely botanical treasures. It is tempting to speculate that he may have been an agent of the Spanish provincial government, ordered to prevent Jussieu from exporting the plants.

Alone and penniless, Jussieu turned again to practicing medicine. Wandering from La Paz to Cuzco, and so to Santa

Cruz and the gold mines at Potosí, he slowly collected again the seeds and bulbs of the Andes, shipping them to France whenever opportunity permitted. At last, in 1755, he reached Lima—only to learn that the greater part of his first collections had been lost at sea. From this heartbreaking experience Jussieu never fully recovered, although he did have one consolation for the twenty years he had spent in Spanish America: the Peruvian heliotrope.

It had reached France in 1757, and had been grown at first in greenhouses. But it proved hardy, and its aroma soon made it a popular bedding plant in the great gardens of the nobles, earning it the familiar name "royal fragrance."

In 1775 the Duc d'Ayen had sent seeds from his famed gardens at Saint-Germain to Phillip Miller in London. A year later, Miller reported, the plants had flowered, and their seeds had ripened successfully. Thus introduced into England, the Andean heliotrope, with its vanilla scent, eventually became a favorite flower in Victorian gardens, where it was often given unlikely names, such as "cherry pie" or "Miss Nightingale" after that heroine of the Crimean War. The prim, pale flower found its way into sachets, and into perfumes and other cosmetics, and its subtle yet penetrating scent became characteristic of the period.

Peruvian heliotrope came late to the United States, but by 1850, as an import from English seed houses, it was growing here and there in gardens in the eastern states.

Garden Heliotrope

Garden heliotrope, or *Valeriana officinalis*, is a far less dramatic flower. A native of the Near East and southern Europe, it grew in ancient Egypt and was known in Greece and Rome. Its name is derived from two Greek words—*helios*, sun, and *trepein*, to turn—because it was mistakenly believed capable of turning its head with the sun.

Throughout history magic powers were ascribed to

valerian, which like many ancient herbs absorbed its power from the will of the user. Legends tell us that Circe and Hecate, for instance, employed it for casting spells and to attain prophetic vision. It was also considered a beneficent herb that offered protection from evil.

In the Middle Ages it made its way into monastery gardens. Albertus Magnus, who grew it at Cologne, called it "turnsole"; Gerard explains that the name refers not to its ability to follow the sun but to its blooming at the summer solstice, when the sun itself once again begins to trace a diminishing arc across the sky.

The plant had reached England by about the ninth century, and eventually grew in cottage herb gardens. It was used as a cure-all for fevers, coughs, and "histerick fits," and also to flavor wines and other enticing drinks. In rural England it is called "setwall," and is still used as a sedative for nerves and whooping cough. A limited amount is grown commercially in Shropshire, and also in the Netherlands.

Garden Heliotrope in America

It was also for flavoring and for medicinal use that garden heliotrope was grown in New England. At Williamsburg it bloomed beside the larkspur and the Canterbury bells; the Quakers grew it in Philadelphia; and the Shakers, America's first commercial herb gardeners, cultivated it for decades for the products they sold so widely.

The root of the plant possesses the herbal properties, and it is dug in the fall and dried. According to M. W. Kamm's popular *Old-time Herbs for Northern Gardens*, first published in 1938, garden heliotrope was still being grown commercially in New York State and Vermont at the time. It is widely grown today in Japan, where its roots continue to be used medicinally and as a relief for nervous disorders.

Garden heliotrope became a wayside weed in Europe long ago, and it escaped from early American gardens as well.

For two centuries it has grown so thickly about some eastern ports, particularly in Virginia, that it is thought to have come from Europe in the ballast of sailing ships.

The perennial red heliotrope, *Valeriana officinalis rubra*, also grown in American gardens, is a latecomer from Persia. It is often called "Jupiter's beard" for its long blooming racemes of deep-red flowers, but it has no history as a medicine or food.

Garden heliotrope is available today in striking hybrids, four feet tall, with cut-leafed foliage and flat umbels of flowers in shades ranging from white to purple that bloom from June to frost. These are irresistible traits, but only Peruvian heliotrope can fill the garden on warm summer evenings with an intoxicating fragrance. The originally small, shrubby plant with pale-violet flowers has undergone extensive hybridization, beginning in the late nineteenth century. By now it is available not only as a semidwarf, but also as a tall, graceful plant that may attain a height of thirty inches. Both varieties, grown as tender annuals in many American gardens, bear wide flower heads in shades ranging from pale blue to dark violet. And both retain the delicious perfume of the South American wildflower—a haunting memorial to Jussieu, the lonely plant hunter of the Andes.

HOLLYHOCK

Behind the fantasy of Shakespeare's *A Midsummer Night's Dream* lie centuries of folklore, such as this old English recipe:

TO SEE A FAIRIE
Buds of hollihocke
Topes of Thyme
Flowers of Hazel

In Rose-water brine
Steeped in the sun
Rubbed on the eye
On a moonlight midnight
A Fairie ye'll spy.

By the Elizabethan era the hollyhock was universally grown in England. Shakespeare knew it, and so did Lord Burghley, lord high treasurer of the realm, whose garden at Theobald, near London, was said to be the fairest in England. It was supervised by John Gerard, from whom we learn that the first double hollyhocks grew there.

But hollyhocks are not English flowers, and although they were first brought to Britain by the Romans as an essential and easily grown food and medicine, they were not Roman either. They originated in China and India, and like various other plants mentioned in this book had been carried over the old caravan trails to the eastern Mediterranean because of their general utility. After Rome departed from Britain they seem to have been lost for a time, but they came back during the Middle Ages and were flourishing again when Elizabeth I ascended the throne.

The Crusaders

The Crusaders, returning to England from the Holy Land, brought back many seeds with them, including those of the "mallow" or hollyhock, which served many needs in the Holy Land. It was traditionally grown and eaten as a pot herb—Job laments that he was despised for his poverty and in his need to eat "cut-up mallow for meate"—and the Crusaders found that it was an excellent insect repellent. In those days before carpets, floors everywhere were strewn with straw and herbs. Hollyhock leaves, both fresh and dried, were scattered as a top layer and effectively controlled the lice, fleas, and bedbugs that commonly infested floors and beds, and the clothing of rich and poor alike.

The cultivation of hollyhocks was given a new impetus

in the sixteenth century when the first great wave of Huguenots arrived in England, fleeing religious persecution in France. They brought with them a vital new interest in horticulture—new techniques, new plants, and new hollyhocks, including a double yellow one that had been developed by the Duke of Orléans, and a double purple that Gerard reported as being grown almost everywhere and "called by diverse people the Outlandish Rose."

Chinese Hollyhocks

Just as England was becoming used to these French hollyhocks, the first cargoes of plants from the Orient were arriving in London. Among them were hollyhock seeds.

In China, we are told, the hollyhock had been cultivated as a "precious herb" even before written records. Its tall spires of bright double flowers were grown as ornamentals in mandarins' high-walled gardens, and for food in the gardens of peasants, who ate its leaves as a common green and its flower buds as a special delicacy. The mucilaginous juice of the stem and roots was used medicinally for coughs, insect bites, and bee stings. Hollyhocks were also planted near beehives because, from their nectar, bees made quantities of honey in an age when other forms of sugar were unknown to most of the world.

The Chinese hollyhocks that arrived in England produced a large double flower in a wider range of bright colors than Europe had ever seen before. With its manifold usefulness and its beauty to recommend it, the hollyhock became irresistible to Europeans.

Hollyhocks in America

The first record of the hollyhock's introduction to New England is a bill dated 1631, from a London grocer, for half an ounce of "holyhocke" seeds—price, twopence—purchased by John Winthrop, Jr., at Governor Winthrop's direction.

Hollyhocks flourished in New England thereafter, and in

time were grown from Maine to Georgia. They were not the ornamental Chinese doubles, but the old reddish flower, *Althea rosea;* and they were used, as they had been for generations in Europe, for pot herbs, insect control, poultices, and the relief of chilblains. Colonists also brought in the marsh mallow (*A. officinalis*), a close relative of the hollyhock, whose medicinal value, like that of *A. rosea,* resides in the mucilaginous juice of the leaves and roots. When boiled and mixed with honey, they produce a delicious and effective syrup that relieves sore throats, bronchitis, colds, and ailments of the digestive tract.

As modern medicine crept in, home remedies and physic gardens were abandoned. Today the mallow has escaped to the marshes that give it its name, and many hollyhocks grow wild on country roads.

Commercial Use of Hollyhocks

As commerce and manufacturing soared in Britain, textile factories sprang up and there was a growing demand for dye. An old rumor claimed that hollyhocks would produce an excellent blue dye, equal to indigo. Experiments were so successful that acres in England were planted to hollyhocks, and by 1821 their dye was widely used.

Further investigation suggested that the hollyhock might also be a source for textiles. It belongs to the cotton family, and perhaps its strong fibers could be used for hemp or as a substitute for flax. From the hollyhocks grown in an experimental plot of 250 acres, textiles were in fact successfully manufactured; but the cost proved to be prohibitive and the project was abandoned.

Hollyhocks in Art

The first Europeans in the East marveled at the exquisite art of the Orient and the subtle use of flowers in the designs of textiles, silks, and embroideries. The simple beauty of the hollyhock motif appeared in India, China, and Japan. In England, Sir Joshua Reynolds incorporated hollyhocks as an atmospheric touch in a number of his portraits, and later the Impressionists were fond of painting them.

Of course they found a place in the herbals. A colored etching of a hollyhock by John Miller in *An Illustration of the Sexual System of Linnaeus*, which was published in 1777, caused Linnaeus himself to exclaim that it was more beautiful and more accurate than any since the world began.

Botanical Bits

The hollyhock belongs to the Malvaceae family, of which there are about 1,000 members, all of them benign, all of them mucilaginous. Many of them are little-known tropical and subtropical plants.

The history of the hollyhock is inherent in its name. The mallows were generally known as "hocs" in Anglo-Saxon days. The reverence for their healing powers is found in the folk name "holig-hoc," or "holy hoc," eventually corrupted to "hollyhock." The generic name, *Althea rosea*, also refers to its medicinal qualities. *Althaia* is Greek for "healer," and records show that altheas were used for medicine in Greece 200 years before Christ.

The hollyhock can be grown as an annual, biennial, or perennial. The original flowers were simple in form and muted in color, five-petaled and stemless, growing on tall unbranched spikes; and the plants were cultivated, as we have seen, for "meate and medicine." The elegant modern hybrids, with bright-colored four- and five-inch flowers that are double, semidouble, or pompons, and grown entirely as ornamentals, would scarcely be recognized by early gardeners.

Our immigrant hollyhock has only one enemy, a foreigner: a fungus from Chile that came to the United States in 1875. It creates rust spots on leaves and stems, and eventually causes the leaves to turn completely brown and fall off. This fungus can, however, be controlled by persistent spraying, and the magnificent hollyhock continues to hold a high place in our gardens.

Hollyhock cheeses, the round seed cases neatly packed with delicious seeds, are not the least of its treasures. Down the ages children in gardens of China, Egypt, Europe, and America have chewed the seeds of hollyhock cheeses with wonder and delight.

HONESTY

The Puritans called it "honesty" and took it to Massachusetts and planted it in their first gardens. Why? It was not grown for food, nor was it an herb for healing ills or seasoning food. It added nothing whatever to the welfare of the colony.

There seems but one excuse for such worldly indulgence—that honesty sustained homesick hearts through the first bitter winters. Bouquets of its silvery seed pods decorated mantels and corner cupboards—nostalgic symbols of former gaiety. Vanity? Perhaps. But generations have smiled and noted ironically that the only seed the Pilgrims brought to New England was honesty.

The sentiment with which it was regarded is conveyed by the folk names that still cling to it. Some sound mercenary: "silver penny," "silver shilling," "moneywort," "money plant," "money-in-the-pocket," "pennyflower," and "money seed." Others are more descriptive: "white satin," "satin seed," "satinpod." But honesty acquired an older name—"prick-song flower," which suggests song fests in early English homes—from the needle-sharp point on each seed pod, which was once used to prick out the notes of songs on thin paper, a common practice before music was printed.

There were other names for honesty, too—names from

ancient days, which had more sinister connotations: "Judas pence," "Merlin's money," and "moonwort." It was listed by old English herbals as a plant for casting spells, and in France, where it also bore this reputation, it was known as *herbe de lunette*. Perhaps the supernatural power assigned to honesty derives from a confusion of names in the Middle Ages: a poisonous fern used by witches and wizards before the church limited their reign was called moonwort, too. In any event mystery clung to the flower with the strange seed pod, and from this association it acquired its official name: *Lunaria*, or moon flower.

Lunaria is a Eurasian plant, whose date of introduction to England is uncertain. The annual form, *Lunaria annua*, grew there for generations. William Aiton, of Kew Gardens and author of *Hortus Kewensis*, says the biennial, *L. biennis*, was brought in from Germany in 1570. The perennial, *L. rediviva*, arrived a short time later. Both species are easily propagated by roots and seeds and adapt to a variety of soils and climates.

The lunarias belong to the mustard family and, like all crucifers, are benign. Their roots were used in England and on the Continent for salads, but there is no record that they were ever cultivated for food in America.

Lunaria in America

Governor Winthrop, acutely aware of the part gardens would play in the success of his colony, carried a copy of Gerard's *Herball* with him to Boston. Gerard had much to say on *Lunaria*, and his enthusiasm and the plant's popularity helped Winthrop overlook the frivolity of growing it in Massachusetts. It was soon cultivated in gardens throughout New England. A Cape Cod diary of the eighteenth century mentions "bayberry and honestie gathered today for bouquets in winter parlor." In southern Connecticut, honesty quickly escaped the discipline of gardens, as we learn from Timothy Dwight. An army chaplain in the Revolution, later rector of

Greenfield School, eventually president of Yale, Dwight rode
on horseback about the colonies gathering material for a book
published as *Travels in New-England and New-York*, which
records, along with his observations and criticisms of agricultural practice, the fact that honesty was growing by then as a
roadside weed.

The earliest advertisement for honesty appears in a Boston paper in 1771, with this added note: "Available only in
small quantities that all might have some."

But in the rush of new flowers that poured into America
in the nineteenth and twentieth centuries, the charm of the
old unusual plant was almost forgotten by gardeners.

Seeds More Appealing than Flowers

Honesty has always been grown for its seed pods rather than
its flowers, which are numerous, fragrant, and purple. It is a
handsome plant eighteen to thirty inches tall, with coarse,
toothed leaves. The spring flowers are replaced by thin flat
seed pods in late summer. When the pods are dried and the
outer shells removed, an inner transparent shell is revealed.
Through this shining purselike disk the dark seeds may be
seen: hence the name "honesty." The pods are about an inch
and a half in diameter, and suggest silver pennies. Once upon
a time children coveted them and carried them in their
pockets for days—an ordeal the fragile-seeming seeds withstood with surprising robustness.

It is this characteristic that has made honesty so popular
in winter bouquets. The shimmering transparent pods complement the red hips of *Rosa rugosa*, brilliant pyracantha
sprays, or Japanese lantern pods, all immigrant plants introduced to America long after honesty grew in Puritan gardens.

Its seeds spread freely, so it may still be found growing
wild in a few localities, but it is now grown chiefly for commercial florists, who use it as an exotic element in winter
flower arrangements.

HORSE CHESTNUT

The horse chestnut arrived in America at the height of that expansive colonial period when estate management and horticulture were a gentleman's obligation, and the introduction of a new tree from abroad was a subject for ardent conversations, copious correspondence, and detailed plantation notes.

John Custis, the dean of Virginia gardeners, whose daughter-in-law became known to history as Martha Washington, recorded the first planting of a horse-chestnut tree in America. He had received the seed from Peter Collinson, of London, and the event took place on the tidewater Custis estate in 1736. Sadly enough, a later note reveals that the young tree did not survive.

It was not until 1763 that John Bartram, who had also received his seed from Peter Collinson, could report that a horse-chestnut tree was actually blooming in his garden on the Schuylkill. Once established, this splendid tree, with its great clusters of flowers, won wide acclaim, and other gardeners eagerly sought horse-chestnut seed. After the Revolution the trees were grown as ornamentals on plantations throughout the south, whose climate they found congenial. Gradually they spread northward into all the Atlantic states, and they followed the pioneers to the west, where they became a symbol of progress, showing that the settlers could afford to plant for beauty and shade and that their towns were permanent and prosperous.

Origin and History

The horse chestnut still grows wild in a few isolated areas of the Balkans, to which it is native. It is said to have survived since the Tertiary period.

Before the fifteenth century it was well established as an ornamental in Turkey. Ogier Ghiselin de Busbecq, the Austrian emperor's ambassador to the Turkish court and an amateur botanist, saw these great flowering trees in the sultan's garden at Constantinople around 1550, and sent some of their seeds—along with lilacs, mock orange, and tulips—to his botanist friend Pietro Mattioli, an Italian serving as physician to the emperor. Busbecq's horse chestnuts were planted in the royal gardens, where Charles de Lécluse, the Flemish botanist, saw and greatly admired them. The emperor gave him seeds, and it was through Clusius that the horse chestnut, as well as other plants from Turkey and the Levant, was introduced to the great gardens of Western Europe. By 1633 they had reached England, and were well established there when Peter Collinson sent seed to American gardens a century later.

Derivation of a Name?

In Turkey this ancient tree has always been called "at-kastane," of which "horse chestnut" is a literal translation. Tradition says the Turks gave it the name because the nuts were fed to horses as a physic and to relieve them of short wind. Mattioli translated the Turkish directly into Latin, and *Castanea equinna*, or, later, *Aesculus hippocastanum*, became the botanical name for the tree in the Western world. American botanists, incidentally, claim the name "horse chestnut" is derived from the leaf scars on the branches: they form a perfect horseshoe, the bundle traces appearing as the nail heads.

It is an oddity that today the horse chestnut is commonly called simply "chestnut," despite the fact that the true chestnut is a quite different tree.

Botanical Bits

The horse chestnut has one of the most complex flowers of any tree. The showy erect clusters, measuring from eight to fifteen inches, bloom in May or June. The red-tongued white flowers have yellow spots that guide the bees and other insects beneath sprays of pollen-topped stamens to the nectar at their base, thereby ensuring pollination.

The leaves are compound, with seven leaflets from five to nine inches long. The trees, which may reach a height of 100 feet, form a canopy of dense shade. There are twenty-five species growing in Asia, Europe, and America today, and pictures of them appear in innumerable flower books. The rarest of these, painted by J. N. Mayrhoffer, was one of the illustrations in a flower book dedicated to the crown princess of Bavaria and published in 1816. Detailed and accurate, it is as richly elegant as the tree itself.

In the United States our native buckeyes, best known in California and Ohio, are closely related to the horse chestnut. But these are less dramatic trees, with smaller compound leaves of five leaflets and less showy flowers, which bloom in midsummer. They attain a height of no more than forty feet. Like many other immigrant plants, the European horse chestnut now exists in far greater numbers in America than the native buckeye.

HYDRANGEA

A century ago hydrangeas were the most popular shrubs in America. They were the perfect complement to the architecture typifying that era of growth and expansion. When houses boasted elaborate front porches, bay windows, decorative eaves and turrets, and a rich encrustation of gingerbread, an ornamental hydrangea from the Orient was considered the choicest planting for the front yard.

The *Hydrangea paniculata grandiflora* was an impressive bush. In the late summer it was prolific, with great, drooping heads of snowy blooms; and, amazingly, these endured through the fall and into the winter, dramatically changing from white to pink and then, with the first snowfall, to greens and purples. To possess at least one hydrangea became the ambition of every homeowner with a twenty-foot lot, and America was soon calling it "old pee-gee."

Many towns still boast front yards graced with hydrangeas, now expansive, ornate bushes, which symbolize the history of the community at the turn of the century.

Hydrangeas in Europe

Engelbert Kaempfer was a young German doctor who was caught up in the great wave of botanical enthusiasm that swept over Europe and through the universities in the seventeenth century. Just after he finished medical school in Westphalia he was offered the post of secretary to the Swedish ambassador to Persia. The botanical riches of the Near East so excited him that he determined to pursue his avocation further.

Upon returning to Europe, he immediately enlisted as surgeon to a Dutch East India Company fleet about to sail to the Orient. He spent the next ten years, wherever the fleet put in to harbor, collecting, drawing, and writing about the plants he found. He arrived in Japan in 1690, and, though prevented from collecting there, he was an especially interested observer of the flowers he found.

In 1712 he published his *Amoenitates Exoticae*, or *The Charms of the Foreign*, the first detailed account of the history and habitat of Oriental plants and their medical and economic uses, which he illustrated with exquisite and accurate watercolors. In it he included the first description of a hydrangea to reach Europe. It described many other Oriental plants then unknown to American gardens but since grown familiar, including camellias, azaleas, peonies, and tiger lilies.

P. F. Siebold, another young German doctor, followed

Kaempfer to Japan for the Dutch East India Company. He, too, became an ardent botanist and collector, and in the course of a half century spent in the East he sent hundreds of plants to the Leyden Nursery, which he had helped to establish in Holland. Among them were the first hydrangeas to reach Europe.

Carl Maximowicz, a young Russian botanist and friend of Siebold in Nagasaki, introduced hydrangeas to Saint Petersburg. His enthusiasm soon made him the first European authority on their cultivation and on Japanese flora in general.

These men demonstrated to the Japanese the commercial value of their plants, and did much to break down their policy of isolation from the Western world, and to prepare the way for the peaceful reception of Commodore Perry and the American fleet that sailed into Tokyo Bay in 1853.

Hydrangeas in America

Charles Wright, an American botanist, accompanied Perry, with orders to observe and report on Japanese flora. He was not authorized to collect plants himself, but his enthusiasm for the beauty and variety of Japanese flowers persuaded George Rogers Hall, a young American doctor practicing in Yokohama, to become a collector. In the next decades Dr. Hall introduced many Eastern shrubs and flowers to American gardens, among them the Japanese wisteria, magnolias, and the *Hydrangea paniculata grandiflora*, now so familiar in old front yards.

The hydrangeas arrived in America in the midst of our era of expansion. In the east, manufacturing towns were growing in size and wealth. In the west, free land was being settled by pioneers. Early settlers had depended largely on European nurseries for their shrubs, bulbs, and seeds, but our rapidly growing nation demanded greater and more personal service.

To answer the demand there appeared the plant peddlers,

who traveled on foot or horseback through towns and villages and the backwoods from Maine to the Ohio valley. Some had catalogues offering a few dozen trees and plants from some eastern nursery, but most carried only a bundle of one popular shrub, which they sold directly to customers in villages and on farms. Among their most popular wares was the hydrangea. Thoreau, describing Concord in his youth, says that after the plant peddler's visit every Concord yard, whether large or small, bloomed in the most democratic manner with the same shrub. Invariably this was *H. paniculata grandiflora*.

Hydrangeas maintained their popularity, and in 1900 the Arnold Arboretum at Harvard University sent E. H. Wilson to the Far East to search for more varieties. He was an Englishman who had joined the Arboretum staff in 1906 as a young man, and became one of the great modern plant hunters. He spent so many years in the East, nineteen of them, that he came to be known as "Chinese" Wilson. He introduced more than a thousand new plants.

On a shore in Japan he found, growing wild, the *H. macrophylla*, which had been discovered by Charles Maries in 1880 and introduced to America via England. It was soon imported in quantity directly from Japan, and its beauty made it a rival of old pee-gee. Another hydrangea that Wilson found, growing wild in China, he named the *H. sargentiana*, for the director of the Arnold Arboretum, but it never became generally popular.

We do have native hydrangeas and a word must be said for them. *H. arborescens* is found along the streams and in the mountains from Pennsylvania to Florida and westward to the Mississippi. The rather similar *H. cinerea* is more numerous in the south, as is the climbing hydrangea, *H. petiolarus*, with its twining stems eight to twenty-five feet high.

In 1925, H. F. Comber, an English plant hunter exploring in Chile, discovered a new species, the *H. intergerrima*, growing in the Andes.

The Blue Hydrangeas

The stellata hydrangeas, with their blue flower heads, still popular in our own day, are the result of abnormal amounts of iron or alum in the soil where they are grown—sometimes introduced deliberately. They were originally made popular by Cecil Rhodes in South Africa, and were known as "Cape Town hydrangeas." In America they were a feature of the elaborate gardens of Newport in its heyday, and they continue to thrive in the salt air of many seaside resorts. What would Cape Cod dooryards be without their bright-blue hydrangeas?

Botanical Bits

The hydrangeas belong to the Saxifragaceae family, of which the familiar mock orange is also a member. They are ancient Asian plants that Europeans presumed to rename in Greek and Latin, "hydrangea" being derived from the Greek words *hydr*, water, and *angeion*, a vase, referring to the watery habitat of many wild hydrangeas and to the vaselike shape of the seed capsule. They have opposite leaves and compound flower heads of various shapes, whose marginal flowers are usually sterile and showy. Hydrangeas may be propagated by summer cuttings, and an enormous number of varieties has been offered by nurseries in the past.

The best known of the Oriental hydrangeas are the paniculatas, the macrophyllas or hortensias, and the stellatas—the blue hydrangeas.

Although hydrangeas are seldom found adorning modern model cities and their subdivisions, or the grounds of condominiums or multiple-housing projects, they are still planted by householders with a sense of nostalgia and an admiration for their enduring beauty. In 1963 botanist, author, editor Norman Taylor wrote that *Hydrangea paniculata grandiflora*, old pee-gee, "is still the most widely cultivated shrub in the country."

IRIS

When King Thutmose III of Egypt conquered Syria in 1479 B.C., he had on his staff a botanist whose mission was to search for new plants. The king was a gardener at heart, and he coveted the flowers of Syria as some conquerors covet gold. So, after he'd won his victory, he had his camels laden with iris roots and rosebushes, and the bulbs of tulips and crocus and lilies, all of which were carried back to Egypt and planted in the royal gardens. Then it occurred to him that nothing could better represent his conquest of Syria than sculptures of the captured flowers, incorporated in the walls of the temple of Amon at Karnak. And there they are today: tulip, crocus, lily, rose—and iris.

During the Crusades, Louis VII of France found Thutmose's iris growing in Egypt, and was charmed by it. He, too, adopted it as a symbol of conquest, and had an iris carved as a device on his coat of arms. It became known as the "fleur-de-Louis," a term later corrupted to "fleur-de-luce" and then to "fleur-de-lis."

This true iris of the Mediterranean area is not easy to grow in our climate. The familiar irises of our gardens are hybrids, developed much later, principally in Dutch nurseries, and they are hardier and flower earlier than the Mediterranean iris.

Source of Iris

The iris, which is found only in the Northern Hemisphere, probably once grew wild over an area that circled the globe and reached from the Arctic Circle to North Africa. Traces of its northern limits have been found in Alaska, in the northern Saint Lawrence region and the Gaspé Peninsula, and across the Pacific in northeast Asia. The ice age may have removed it from Europe and pushed it to North Africa's Mediterranean littoral—which is as far south as it has ever traveled. It will not thrive in the tropics, and has reached perfection only in the North Temperate Zone.

As the Romans pushed their roads into Egypt and Syria, so that they could transport wheat and olive oil to Mediterranean ports, they coveted the iris that grew there so abundantly. Iris roots, along with the wheat and the oil, soon found themselves on Roman ships, and planted in Roman gardens around the northern shores of the Mediterranean, where they flourished as happily as in North Africa. Among their admirers was Virgil, who noted the infinite variations of reds and blues and yellows of the iris in his garden, and wrote that it was aptly named after the Greek goddess of the rainbow.

Centuries passed. With the fall of the Roman Empire and

the abandonment of great estates, many cultivated plants were lost to Europe, iris among them. But when the Moors of North Africa invaded Spain, they brought with them a rich endowment of medicinal herbs and cultivated plants, including iris.

For 800 years these flowers throve in the walled gardens of Moorish castles. Then, in 1492, Ferdinand and Isabella drove the last of the Moors from Spain, and the plants that had been so long immured flowed out across the land to the ultimate enrichment of all Europe. The close political and economic relationship of Spain and Portugal to the Low Countries brought the iris to Holland, whose gifted gardeners translated it into the ancestors of the Dutch, Spanish, German, and English irises of our own day.

Although the Romans had undoubtedly carried the iris to England, it had been neglected there, and was ultimately lost. With the Renaissance came a new vital interest in gardens, in both ornamentals and useful plants, and in the Elizabethan era horticulture became an obligation of the nobility. Iris flourished in the English climate, and soon grew in monastery, castle, and cottage gardens. Later it escaped to the swamps and naturalized itself.

Iris in America

Iris arrived in Virginia with the earliest colonists, perhaps brought in by the first women, who arrived in 1619. Whoever brought it, the Persian iris was well established when Williamsburg was planned, and was on the list of flowers in the governor's first garden. It was later imported to all the colonies, from Maine to Georgia, and throve wherever it was planted.

The only native iris the colonists found was the wild blue one, which grew abundantly in the marshes and creeks along the East Coast. They called it "blue flag" because it reminded them of the yellow flag, *Iris pseudacorus,* that grew wild in

English swamps. For centuries the blue flag was considered the only wild American iris, but as the country was settled a native dwarf iris was found in the Appalachian Mountains. A much more recent discovery is a native Oregon iris, also a dwarf, which blooms in an enchanting array of colors from purple to pink to pure white. The Oregon iris can easily be grown from seed, and its natural hybrids are exquisite: they are a new boon to iris breeders.

One of the most exciting events in the American iris story was botanists' discovery of the Louisiana iris. The swamps around New Orleans had long been gay each spring with tall bearded iris, but no one thought of them as natives; it was assumed that they had escaped from old plantation gardens. When modern New Orleans began to grow, the swamps were drained for building sites and this iris became an endangered species. Many citizens flew to save it, and introduced it into their own gardens, where it adapted quickly. Garden-club members and botanists became interested; the Louisiana swamp iris was collected and widely described; and floriculturists discovered that it was a native after all. No one knows how many varieties there are because it has hybridized naturally for years. In any event, the iris swamps of southern Louisiana were looked at with new eyes. Demands for these brilliant flowers—of unusual size, and in all shades of the rainbow—began pouring in from every state, from Europe, and even from Australia, New Zealand, and the Far East.

Another example of an American botanical invasion abroad is the recent adventure of our native blue flag. When England became industrialized in the last century, she increasingly drained the swamps and creeks that were the home of the native yellow flags. As these began to die out under the pressures of civilization, English conservationists imported the sturdier American blue flags, and in 1890 planted an experimental patch of them at Ullswater, in the Lake District. By 1953 they had filled all the nearby marshes and moved on to

far more distant areas, even crowding out the remaining
native yellow iris in Epping Forest, near London. Today the
American blue flag waves gaily over acres where once the
English yellow iris flourished.

Iris Imports

But long before "American iris" meant more than just wild
swamp iris, foreign iris filled American gardens and took
prizes at county fairs.

The many varieties of European rhizome iris that originated about the Mediterranean are best known to American
gardeners. Behind them lies a long history of travel and experimentation. From the Caucasus Mountains, and from Austria
and Hungary, came *Iris pumila*. Other rhizome iris came from
southern France, Italy, and the Balkans. The crossing of the
Mediterranean and East European iris produced the tall
bearded iris, widely planted in gardens of yesteryear, whose
ease of cultivation, endurance, and adaptability have maintained its popularity. It is often called "German iris," but this
is only a general trade name for a hybrid, and one that did not
originate in Germany.

Siberian iris was introduced into England in 1596, and
the small fluttering flowers perched along its stems gave it the
common name "butterfly iris." For whatever reasons, it was
not generally grown in early American gardens; nor was
Spanish bulbous iris, which Gerard had grown in his English
garden as early as 1633.

We have the great plant hunters of this century—E. H.
Wilson, Frank Kingdom-Ward, and Michael Foster, a leading
authority on their cultivation and classification—to thank for
the wonderful iris from the Far East that bloom in today's
American gardens. Siberian iris from eastern Asia, Korea, and
Manchuria is a relatively recent arrival, as are the true Persian
iris from India and Turkestan and the crested iris from Japan.
These iris and their hybrids, so familiar in current cata-

logues, gardens, and flower shows, owe their presence to
the search for, and planned introduction of, foreign flowers
by our government and our botanical gardens.

Until fairly recently immigrant iris, like most botanical
imports, arrived in our gardens by way of European nurserymen. But today we have a rising generation of American
nurserymen and orchardists, and American breeders now
hybridize and raise "new" iris from seed, with exciting results.
Their nurseries seem to be divided into two specific groups:
those that grow iris for gardens of the north and northeast,
and those that grow iris suited to California and the southwest. Many California breeders specialize in the English
bulbous iris, which is really native to the Pyrenees, and the
tall bearded iris from Kashmir. Commercial eastern iris gardens, which now flourish in New England, the Great Lakes
area, and along the Potomac, specialize in the Dutch iris, so
long exported by Holland nurseries.

Many American iris breeders are amateurs. For these enthusiasts, and for thousands of other American gardeners, the
American Iris Society puts out a quarterly bulletin that is a
source of valid, comprehensive iris lore. In England, the Iris
Society publishes a yearbook of comparable material. But
there are few significant books on iris. Perhaps W. R. Dykes's
1913 monograph *The Genus Iris* is the best known and most
comprehensive work on the subject. *Iris for Every Garden*,
by Sidney Mitchell, is an excellent standard reference, but
even Mitchell deplores the scarcity of authentic iris information and asserts that this was his reason for writing the book.

Commercial Iris

The Romans, the Egyptians, and the Moors thought of iris as
a medicinal herb, and the Romans undoubtedly brought it to
England as such. It was grown for that reason in monastery
gardens, and even today is used in China as a purgative. But its
chief contemporary use is in cosmetics, in the form of orris

powder, which is derived from the violet-scented dried roots of iris. Originally iris root was grown in Florence for cosmetics for the nobility, but today cosmetics are considered a personal necessity by virtually everyone and have become international big business. To supply an ever-expanding market, *Iris florentina* is now grown in the warm fields of Mexico as well as in France, the world's cosmetic capital. Tons of iris root are shipped to France from Mexico each year and are transformed into the cosmetics that are among the chief French exports.

Swan Lake Iris

America's greatest collection of iris is at Swan Lake, in Sumter, South Carolina. This great iris garden, which covers 100 acres, began on a piece of abandoned land when some rejected Japanese iris were dumped there along with other garden debris. The discarded iris sprang to life, spread over adjacent fields, and bloomed so prodigiously that the phenomenon came to the attention of the Brooklyn Botanical Garden. At its request, the original owner of the land made an experimental planting of some iris imports from Japan. These throve even more spectacularly, and the fields became a riot of iris blossoms whose beauty and variety attracted increasing numbers of garden enthusiasts. Eventually Sumter took over the spontaneous gardens as a community project, imported additional iris from Japan, extended the plantings, and opened the tract to the public.

The exhibit is at its height from May 20 to June 15 each year. Among the 200 varieties growing at Swan Lake, one may see iris from the ends of the earth—including a prize Japanese variety that stands five feet high and bears blossoms ten inches in diameter. It is as if Iris herself had chosen this American garden for the end of her rainbow. The iris of many nations that flourish there point the way to a superior world created by international co-operation.

JONQUIL, DAFFODIL, NARCISSUS

"All the world laughs with their glittering." This is the song Demeter, the goddess of agriculture, sang as she rejoiced in the fields of daffodils one Grecian spring. Perhaps it was the very field where Narcissus had gazed at his own beauty in a brook. Perhaps the dancing daffodils, reflected in the stream, suggested Narcissus' punishment for his persistent vanity: the gods transformed him into a flower so he might live beside the brook and gaze forever at his own beauty.

It is such old myths that anchor the origin of the narcissus to the Mediterranean area: versions of them are found not only in Greece and Rome, but also in Arabia, Egypt, Spain, and Portugal.

Long after Demeter ascended Mount Olympus, the Romans carried narcissus bulbs to England; and eventually English men and women took them to North America, Australia, New Zealand, and the East. Few flowers have enjoyed such constant favor and participated in the history of so many nations. Through the ages, Sophocles, Homer, Mohammed, and Shakespeare have sung their beauty.

First Aid for Gladiators and Warriors

The narcissus, wrote Galen, the progenitor of modern medicine, has such qualities that its slimy juices will glue together great wounds, cuts, and gashes. He did not speak without experience, for his first position was as surgeon at the School of Gladiators. His fame soon came to the attention of the emperor, Marcus Aurelius, who made him physician to the Roman legions on the northern frontier. Galen immediately ordered every soldier to carry narcissus bulbs with him as part of his standard equipment, to cleanse his wounds and heal the cuts of sword and spear. As a medicinal herb in the first-aid kit of the Roman soldier, the narcissus was scattered throughout the Roman Empire.

There is little doubt that it was introduced to Britain in this capacity, for there is no record that the Druids knew it before the Romans came to England. During the five centuries of Roman dominance the narcissus escaped from the campsites of the soldiers and the walled gardens of the governors and became naturalized in England's fields, growing wild beside her brooks.

Why These Names

"Narcissus" was the first name by which these flowers were known in the eastern Mediterranean area and North Africa, where they originated. Today *Narcissus* is the generic name of this branch of the Amaryllidaceae family. "Daffodil" and "jonquil" are the names of species of narcissus, of which thirty are recognized, though there are hundreds of strains because of their inherent variability and man's hybridization.

Two twentieth-century authorities, America's Norman Taylor and V. Sackville-West, of England, agree there is no difference among jonquil, daffodil, and narcissus, and that the names are used interchangeably in different localities and countries. But "narcissus" and "jonquil" usually indicate flat-faced flowers, and "daffodil" the trumpet type.

The Roman soldiers sang of narcissus, the flower of asphodel, which had grown in such masses in the meadows of Homer's Greece. A thousand years later the Norman soldiers who conquered England still sang about the flowers of asphodel. But time had corrupted this to "d'affodils." As daffodils, with their golden trumpets they announced spring-time in Britain for five centuries more. But when James I married Anne of Denmark in 1589, she brought her favorite flower from Denmark's royal garden with her. This was the jonquil, a yellow flower with a short trumpet that proved to be almost identical to the wild daffodil of the English countryside.

The jonquils had arrived in Denmark by way of the Low Countries, to which the Spanish had brought them during their long domination of the area. In the Iberian Peninsula the native narcissus had originally been called "juncas," the Latin for rush, because of its leaves and habitat. Through the centuries this had been corrupted in neo-Latin to the Spanish "junquillo" and the French "jonquille." By the sixteenth century jonquils had become naturalized, grew wild in the woodlands, and were cultivated in the royal gardens.

Queen Anne planted her jonquils at Kensington Palace, and she and the court continued to call them by that name; although English countryfolk went on calling the wild narcissus "daffodil." Anne was an exquisite needlewoman, and she wove jonquils into tapestries with beauty and skill. Her ladies, vying for her favor, embroidered and designed dresses and draperies, cushions and carpets, with patterns of jonquils. Ever since, the *Narcissus jonquilla* has been called "Queen Anne's jonquil." (See Queen Anne's lace.)

The popularity of jonquils at the Stuart court resulted in a near tragedy for the wild daffodils in England. The demand for them in the London flower market increased tremendously, and the Tenby gypsies, who supplied the market, stole them by the thousands from the meadows of Pembrokeshire, the only place in England where they grew wild in such quantities. Their depredations aroused public protest, and for the first time in recorded history masses of people demanded legislation for the conservation of wildlife.

Queen Anne's jonquils also precipitated another action new to history. Anne was not herself a gardener, but the age of gardens had begun burgeoning in England during Elizabeth's reign; and, like Elizabeth, Anne was exceedingly fond of display and would brook no rivalry that left her in the shade. With royal dash, she gave a garden party at Kensington Palace when her jonquils were in full bloom, and in the midst of the fete announced that she was establishing Kensington Palace Gardens as the first public gardens in England. With this munificent gesture, she inaugurated the custom of donating private lands for the use of all the people.

The rapid spread of the jonquils' cultivation in English gardens is reflected in the notes of England's early horticulturists. The first herbalist to describe them was William Turner, father of British botany, from whom we learn that when they were planted at Kew Gardens in 1551 only sixteen varieties were known. In 1629 Parkinson reports nearly one hundred varieties, and in 1665 John Rea notes that the varieties of daffodils are beyond his scope, and adds that "they yield countless new diversities." By the eighteenth century Phillip Miller laments that England must import new varieties from the Low Countries because only such gardeners are willing to undertake the three-year labor of producing the new bulbs and thereby reaping wealth. His envy can be understood; one Dutch bulb of a new variety sometimes brought as much as fifty pounds in London. But the commercial growing of narcissus in England did not begin until 1865, when bulb nurseries were opened in the Scilly Islands.

Jonquils in Jamestown

There were neither daffodils nor narcissus on this great continent until women came over from England in 1619 as prospective wives to the men who held the fort at Jamestown. Tradition says that jonquil bulbs were among the garden makings they brought with them stuffed in their apron pockets for safekeeping. The tradition may well be based on fact, because the first daffodils bloomed in Virginia in the spring of 1620 on the plantation of George Mansie, one of the first to be established in the Tidewater. By 1650 tobacco was supporting a small landed class of plantation owners, and jonquils grew in every garden along the James and throughout the Old Dominion.

Jonquils also grew in the first gardens of New Netherland, and in all the early colonies along the coast. In many isolated settlements homesick settlers still clung to the old familiar names by which narcissus had been known in English villages: daffydowndillys, hooped skirts, codlins-in-cream, lenten lilies, butter-and-eggs. Descendants of these daffodils spread across the country with the pioneers, and today, from Long Island and tidewater Virginia to Puget Sound, they add radiance to springtime in America.

Oregon has become the center of our commercial bulb growing, largely through the efforts of Jan de Graff, the descendant of a long line of Dutch bulb growers. Millions of bulbs are cultivated with modern mechanized aid, and many new varieties are developed. Among the thousands of seedlings, a sport or cross occasionally appears that is a triumph.

One such is the noble golden-trumpet King Alfred, which became an outstanding commercial success.

Jonquils in China

In the Orient the narcissus is known as the "sacred lily of China." Its purity and promise are a symbol of the Chinese New Year, and it is widely grown in Chinese homes. The flowering of the sacred lily on New Year's Day, after only a few short weeks in a bowl of water, seems a small miracle. It is also the culmination of three long years of careful nurture, which begins when young bulbils are planted in rich soil and, the following fall, dug, dried, and stored. This process is repeated twice more, so that the bulbs may store up enough food and strength to explode into bloom. At the end of the third year a master forcer watches the dormant bulbs until the first leaves are ready to emerge; then the bulbs go on sale. Placed in a bowl of water, they need no other nourishment to grow rapidly, and bloom.

In America we know this Oriental narcissus as the "paper-white" that blooms at Eastertime in thousands of homes. Originally its bulbs were imported from China, but when Chinese exports were closed to American nurseries the art of forcing was developed by our own seedsmen, and packages of forced domestic narcissus now appear in many a flower shop and supermarket.

The Chinese, strangely enough, look upon the narcissus as a foreign plant, although they have cultivated it for more than a thousand years. A Chinese book of the ninth century says that the Arabs brought "narce" bulbs into the ports of Fukien province. According to other records they were carried to China by Arabian caravans, in a trade that flourished until 1270. And there is one old legend that claims they were brought from Persia by none other than a wandering Chinese.

Narcissus culture in China is still confined to the province of Fukien, where whole villages have practiced the art of bulb forcing for centuries. Sons of the master forcers have developed the exclusive art of bulb carving, which requires great skill and patience, and causes forced bulbs to burst into bloom with curved crooked leaves and flowers shaped into the form of tigers, fairies, and other fantastic creatures. Such bulbs are called crab-claw narcissus and command the highest prices.

The Danger in Daffodils

Although Roman soldiers considered them an antiseptic, and both Galen and Gerard recommend narcissus bulbs for relieving a variety of ills, they are now classified among the poisonous plants of our gardens. They are never eaten by animals, and records show that people who have gathered and eaten young shoots, mistaking them for wild leeks, have suffered dire consequences.

Galen to the contrary, the sticky juice of narcissus can cause dermatitis. Even Pliny the Elder insisted that all was not gold that glittered in the narcissus, and that the very name, derived from the Arabic word *narce*, originally meant numbness and dulling of the senses.

Daffodils do contain toxic properties, but the alkaloids they carry have not been fully identified. Modern science is re-examining the "green medicine" of simpler times, and experiments suggest that compounds extracted from daffodils may prove useful in the control of multiple sclerosis.

Today's Daffodils

The daffodil owes its modern face and place in our gardens to Peter Barr, who became the ultimate authority on all the species and variations of daffodils and jonquils. He published in 1885 *Ye Narcissus*, a monograph that traces the historical and botanical development; sponsored the first daffodil conference; gathered the greatest collection of wild species ever made; and became, indeed, so enamored of daffodils that he set himself the task of rediscovering all the old varieties described by Parkinson in 1629.

He explored the Iberian Peninsula, and reintroduced many kinds of narcissus that had been lost for more than 200 years. He traveled to the United States, Canada, Japan, China, South Africa, New Zealand, and Australia, seeking old and new daffodils. His greatest achievement was the reintroduction of the white trumpet narcissus, which had been lost to cultivation since Parkinson's day. He came upon it in Spain, in 1885, among a dozen other narcissus varieties that had gone wild in the mountains.

The choice open to contemporary gardeners seems all but limitless: flat-cupped narcissus, a modern hybrid, large-petaled and strong-stemmed; poeticus daffodils, with clusters of flowers; trumpet daffodils, long the most familiar but now available in pink as well as yellow and white; a diversity of miniatures for the rock garden, trumpeted, cyclamen-flowered, cluster-flowering, and now a strain called "butter-flies"; and, of course, jonquils forever.

For a drift of golden daffodils, fling handfuls of bulbs across your garden bed and plant them where they fall. This is highly recommended for a happy springtime.

Narcissus, said Mahomet, are food for the soul.

KALE

Have you had your kale today? If not, hurry down to the supermarket. You'll find it among the frozen foods or in cans on the shelves. You'll also find it on the diet counter, and with the baby foods, and, if you're lucky, fresh among the green vegetables.

Why all this sudden enthusiasm for kale, and what is kale anyway?

Kale and collards are leaf cabbages, cabbages that won't head, and they are one of the most easily grown vegetables and one of the richest sources of vitamins and minerals you can buy. This is what modern nutritionists discovered when they began to do research into the diet and health of the rural population of the south and the crowded urban areas of the north. The south came off with a surprisingly high standard of health, and kale got the credit.

Kale and collards, which grow with a minimum of attention in the garden patch, provide a perfect balance for hog meat and corn bread, the standard diet of sharecroppers and the southern poor in general. Kale and collards supply salad greens and pot herbs throughout the year, and frost heightens the flavor by releasing the high sugar content.

Now nutritionists are recommending that this excellent, inexpensive, and readily available vegetable be provided for all northern urban markets, especially in tenement areas, where it is little known or used. They are urging that kale and collards become a universal food, and producers and processors have been quick to respond.

Though kale once grew in the home gardens of many families, it fell victim to the rapid urbanization of the twentieth century. Truck gardens, often hundreds of miles away, have replaced home gardens, and food distributors examine the personality of every vegetable for its shipping qualities, and for its sales appeal in matters of color, size, and shape. Nutrition comes last.

Modern head cabbage, which ships well and appears crisp and green on the vegetable stands, had almost completely replaced kale by 1910. But today's rapid transit, and refrigerated cars and trucks and vegetable counters make possible the re-introduction of this vitamin-rich vegetable.

A Short History of Kale

Kale and collards are different forms of the same plant. They have one botanical name—*Brassica oleracea acephala*—which means "cabbage without a head," but kale is curly and collards have broad smooth leaves. They are the earliest forms of cabbage, and were cultivated centuries before head cabbage was developed. Their origin reaches back to Asia Minor, but exactly when is not known because of the migration of tribes and the merging of old trade routes in the eastern Mediterranean area.

The Greeks grew kale, and Roman gardeners developed a number of varieties. Records from the first century on show that kales were familiar food. They were carried to northern Europe by the Celts, who, before the Roman legions brought order to Europe, frequently raided the Mediterranean area, and recognizable forms of the word "kale" are found in most European languages influenced by the Celts.

The Romans, who lived so long in Britain, appear to have taken kale with them, and the Vikings who settled on Scotland's shores found that kale grew well there. "Kale," which is the Scottish word for this plant, has prevailed in the English language since Anglo-Saxon days.

Head cabbage was not developed until the Middle Ages. Then it gradually became the dominant cabbage of northern Europe, where the long cool nights and low temperatures encouraged it to head, which it could not do in the south. The keeping qualities of head cabbage also meant that it could be stored, and so provide food for the long winters.

Kale in America

Kale and cabbage were first brought to America in 1540 by the French explorer Jacques Cartier, who planted them along the Saint Lawrence to provide a dependable and familiar food for his men through the winter.

Champlain grew cabbages in his first settlement on an island at the mouth of the Saint Croix River in 1604. The plan of the village, the design of the gardens, and a record of the beds where lettuce, sorrel, and cabbages were planted still exist.

A "bill for garden seeds" sent to Governor Winthrop in Boston in 1631 lists coleworts, a general term used for kale and several cabbages. John Josselyn, in his 1672 *New-Eng-*

land's Rarities Discovered, notes cabbages as "growing exceedingly well."

In Virginia, kales and cabbages grew in the plantation gardens of Jefferson, Custis, and Washington, as in every kitchen garden in the south. Another Virginian, the first John Randolph—not to be confused with the more famous John Randolph, of Roanoke, born more than a generation later—describes in his *Treatise on Gardening* the cultivation of kale. His book was of incalculable value to the settlers because it showed them how to raise English vegetables successfully in Virginia's climate. Ironically, perhaps, he was a British loyalist, who abandoned his plantation when the Revolution started, moved to England, and never returned in his lifetime, although his body was brought back to his birthplace in Virginia.

As the American colonies grew, kale acclimatized itself in the south and became a staple food; head cabbage thrived in the northern settlements, where it became a dependable winter green and was preserved as sauerkraut. But the cabbage of colonial days was a loose, ragged affair, often rank and bitter, and the firm white heads we know were not developed until the twentieth century. Red cabbage and curly-leaf savoy were also introduced in the early days, but were little used until the plant wizards tenderized them a few decades ago.

Chinese cabbage, now familiar in American markets, is a form of kale, a descendant of the wild plant, that has been grown and stored in China for 2,000 years. Other kales introduced to our tables were developed in various parts of Europe over the centuries. The oldest of these is cauliflower, once called "kale-fleur," which originated in Turkey, was used in Egypt and Greece in the sixteenth century, and was brought to Holland, where the Dutch developed the version that became so popular in Elizabethan England. Italy changed her bunched kale of the Middle Ages to the cultivated broccoli,

Belgium developed Brussels sprouts from a primitive kale, and Vienna made famous her *kale-rape* as the modern kohlrabi.

Of these, cauliflower has been in America the longest. Records show that it was introduced by Winthrop's Puritans and Penn's Quakers, and in New Netherland and colonial Virginia. The other versions of kale came into our markets with the later influx of immigrants from southern Europe.

A kale with leaves fifteen feet high has recently been developed for cattle feed in Europe, where surplus kale and cabbage have always been utilized for stock food, as they have been in Asia and our own south.

Kale in the Flower Garden

The last word on kale is that it has reached the flower garden as an ornamental. Floral kale or flowering cabbage is a beautiful bright-green rosette of leaves with a coral or white center—an interesting, edible corsage that came from the Orient. Like a subtle postscript to centuries of mundane service, decorative kale provides a neat and novel border for your flower bed, and a gala salad for a modern meal.

LETTUCE

Lettuce, which has been cultivated longer and used more widely than any other vegetable, is probably the most popular grown today.

Its use began before recorded history. Wild lettuce originated along the Mediterranean littoral, probably in Egypt, and it was known as an edible green throughout antiquity. The Hebrews ate lettuce with their paschal lamb, and according to Herodotus the early Persian kings ate lettuce at their feasts.

Lettuce grew in Greek gardens 500 years before Christ was born, but it remained a festal herb, reserved for ritual dishes. Served with saffron and dressed with olive oil it was used at funeral feasts, to which it lent honor and significance. It was also used to celebrate the Fete of Adonis, the coming of spring: pots of it were carried through the streets with banners and songs, and it was displayed in homes and gardens much as we display Easter flowers today.

In ancient Rome, lettuce remained a luxury, reserved for feast days and the tables of the wealthy. Horace tells us that no proper patrician feast began without a salad of lettuce "to relax the alimentary canal" and prepare the body for a surfeit of food. The Romans grew lettuce in the gardens of the villas they built in England, whose cool damp climate suited it perfectly. It became, indeed, one of the few cultivated plants that survived when the Romans departed after five centuries of occupation.

The British historian Holinshed records that through the Middle Ages, when gardens and orchards were abandoned, meat and fish became the principal diet and lettuce and oil were commonly used "to lighten the stomach." But perhaps one reason lettuce survived is that it was considered a medicinal herb. Anglo-Saxon herbals recommended it as a sedative and a "dilutant of animal foods." Its milky juice, mixed with henbane and poppy, induced sleep. A milder mixture, a broth of chicory and lettuce, was also conducive to repose. Lettuce was grown in the early monastery gardens as both a food and a sedative, and a decoction of lettuce, rue, and sorrel was used as an emetic.

Saint Gall grew lettuce at his famous Benedictine abbey in Switzerland. Plans of its ninth-century herb garden, still available, show plots of celery, cabbage, parsley, leeks, and lettuce, and we know from him that lettuce grew in Charlemagne's gardens in France. It grew, indeed, in the kitchen gardens of every medieval castle. The salads served at the tables of knights and wealthy merchants consisted of lettuce and flowers and herbs in season—violet petals, calendulas, and rose petals, the leaves of mint and sage—

dressed with oil and vinegar; they must have made a dish as colorful as it was tasty.

Many Elizabethan herbalists comment on the importance of lettuce, and recommend that it be eaten at mealtime and before indulgence in drink—because, says one of them, "it staieth the vapours that disturb the head and cooleth the hot stomache which some call heart burn."

Lettuce in the New World

Columbus is thought to have brought lettuce seed to the West Indies in 1493, along with peas and beans, to test the ground—a theory borne out by the wide use of lettuce in Haiti as early as 1565. It was grown in Mexico, and in the Portuguese colony of Brazil, before the Pilgrims landed in New England. The French grew lettuce on the Saint Lawrence, the Dutch in New Netherland, the Swedes on the Delaware, and so did all the English colonists along the Atlantic coast. The Swedish botanist Peter Kalm, who traveled widely in America for Linnaeus during the second half of the eighteenth century, describes typical dinners in Quebec, which consisted of soup, several meats, fish, wild pigeon, and fruit, and, as an invariable third course, salad prepared "in the usual manner."

John Randolph says that lettuce "both Roman and Egyptian" flourished in the lovingly planned and tended gardens of Virginia's plantations. (See Kale.) The Shakers' first commercial herb garden, a "Physic Garden for Home Remedies," at Lebanon, New York, included lettuce, along with sage, horehound, aconite, and burdock, as an essential medicinal. By 1789 the Shaker Seed Wagon, laden with "Shaker Seeds for Home Gardens" supplied many rural areas of our growing country, and by 1840 was selling thousands of packets a year—including countless packets of lettuce seed. (See Peach.)

Kiehl's Botanical Handi-Book, which might be called the

descendant of the Shaker seed catalogue, lists, among the herbs, barks, and berries with medicinal value that are still available, acrid lettuce (*Lactuca virosa*), wild lettuce (*L. elongala*), and garden lettuce (*L. sativa*) as having the properties of a diuretic, narcotic, and sedative—and also, we are relieved to discover, as a condiment and salad.

Lettuce's value as a medicinal was recalled during World War II when English hospitals were cut off from their source of the drug lactucarium in southern Germany. Florence Ransom says that a local source was found in the wild lettuce, *L. virosa*, that sprang up spontaneously on London bomb sites and that was then grown in England's herb gardens.

Wild Lettuce in the United States

Blue lettuce, *L. pulchella*, is the only species native to our continent. It grows on the western plains, where its blue flowers color hundreds of acres; but so far no practical use has been found for it.

A weedy invader from Eurasia, an apparent stowaway in a shipment of seeds, is prickly lettuce, *L. serricula*, which has forced its way into virtually every state, and which is a menace to pasture lands, particularly in Wyoming, where cattle develop emphysema from feeding on it. Furthermore, its innumerable seeds and long tough taproot have, to date, made it impossible to eradicate.

Leaf Lettuce and Head Lettuce

There are three main types of edible lettuce: head or cabbage lettuce, leaf lettuce, and romaine or Cos. Both names for this third type, with its long leaves and crisp heart, are of ancient lineage: "romaine" is a corruption of "Roman," and Cos derives from the Greek island of Kos, in the Dodecanese, where it is said to have been grown first. The botanical name for lettuce, *Lactuca*, refers to the milky juice present in its stems. All lettuces are annuals and belong to the Compositae

family. Their seeds, not unlike the dandelion's, have parachutes that carry them to new ground and that have helped
them spread throughout the world.

Common garden lettuce, *L. sativa*—*sativa* is the Latin
word for "cultivated"—is closely related to the wild, loose-
leafed lettuce from which it is derived. All early forms of
cultivated lettuce were also loose-leafed, and today's garden
catalogues offer many varieties of it. Leaf lettuces are ideal for
growing in home gardens, where they produce delicious
edible leaves in forty to forty-eight days. But it is virtually
impossible to store or ship them, and so they are sold, as a
rule, only in local markets, and only in season.

Head lettuce was not known until the Middle Ages,
when, like head cabbage, it was developed in the cooler European countries. It has come into its own in the past century,
during which growing urban populations have established a
demand for fresh vegetables from distant truck gardens. In
1905 it took nine days by rail from the California lettuce
fields to New York; today, vegetables can be shipped by
plane in a matter of hours. But even head lettuce is a highly
perishable item, and retains its goodness under refrigeration
for no more than three weeks.

A half century ago mildew, blight, and rot attacked the
lettuce crop in California. But the ingenuity of hybridists has
produced a resistant strain, and lettuce continues its success
story. Today it is the fourth-most-important food crop in the
United States, as the lettuce strikes have helped to demonstrate: the workers in the lettuce fields of California and the
southwest have fought for decent wages and living conditions, and have won the right to unionize.

And yet iceberg, the most widely known head lettuce,
has little color or odor or flavor, as gourmets increasingly
complain, and no lettuce has much food value. The endurance
and popularity of lettuce through the ages must be regarded,
then, as somewhat mysterious. Yet how should we get along
without it?

LILAC

Springtime in yesterday's New England was called the "lilac tide." Lilacs bloomed in every dooryard, sprang up beside the barn, shaded the well, hung over the springhouse, crowded the lane, and grew beside the picket fence, offering you a fragrant welcome from either side of the front gate. The lilac was inextricably bound up with home life in Puritan New England.

It was no accident that lilacs grew in these front yards. A front yard was a privilege inherited from England's Middle Ages, marking the home of a yeoman and his right to own land. The yeoman signalized his financial standing by planting his forecourt with ornamental shrubs, and the decorative Oriental lilac, still rare in England when Massachusetts was settled, marked the home of a yeoman of substance and culture. New England felt a close affinity for this shrub, which endured the bitter winters and then welcomed the spring with a lavish display of blossoms.

There are many tales about where and when the first lilacs arrived. Some accounts say they were planted as early as 1638, and one romantic tale concerns an English gentleman, Sir Harry Frankland, who had a mistress in New England on whom he lavished every luxury. He even sent her exotic plants, including the still-rare lilac.

One of the earliest substantiated claims is that they were brought to Portsmouth, New Hampshire, in 1695—not from England, but from Persia—by an English sea captain who had decided to settle in that pleasant harbor town. In any event, the Portsmouth lilacs weave their way through the next century of American history. When Benning Wentworth, the colonial governor, built his mansion in Portsmouth in 1750, the old sea captain presented him with some lilac cuttings, which were planted on the mansion's terrace. Years later other cuttings were sent to George Washington, when he returned to Mount Vernon at the close of the Revolution to devote himself, as he briefly believed, to horticulture. As for Governor Wentworth's mansion, in due course it became the first state capitol building, and descendants of the lilacs planted on the terrace are growing there to this day.

Although there are few published accounts of American gardens in the eighteenth century, lilacs are mentioned in estate ledgers, garden records, old diaries, and the letters and notes of famous gardeners in many colonies.

In 1737 Peter Collinson sent John Bartram a bundle of lilac shoots from London, along with a pertinent note of explanation: "As your neighbors in Virginia, in particular Col. Custis at Williamsburg, who has undoubtedly the largest collection [of lilacs] in that country, desired some of me, I thought you might want them too."

Jefferson's Monticello garden book includes a laconic entry—"Apr. 2, 1729. Planted lilacs"—which many claim is the first authentic written record of lilacs in America. An item in Washington's plantation records is only a touch more eloquent: "Feb. 10, 1786, Buds of lilac much swelled and seem ready to unfold."

We know that by the time of the Revolution, Pennsylvania gardens on the Schuylkill, Wissahickon, and Delaware bloomed with lilacs, and an old New York garden book describes a garden in the Dutch tradition, enclosed within the privacy of a high clipped hedge and "gay with laylocks, snowballs, May roses, and Pans bloemen."

Williamsburg became the capital of Virginia in 1699 and the center of elegance in colonial America. In the Governor's Palace there presently arose the tradition of placing a bowl of white lilacs before the portrait of Catherine of Braganza—a gracious symbolic gesture to the queen of Charles II of England. She had brought him, as part of her dowry, nothing less than the port of Bombay and entry to the port of Canton;

this was the wedge that opened the Orient to England, and carried lilacs, with many other floral treasures, to palace gardens in Europe and to the great plantation gardens of the early south.

On a humble level, the plant peddlers of the early nineteenth century played a part in spreading lilacs across America. They followed the pioneers over the Alleghenies into the Ohio valley and the Northwest Territory, and found them eager to purchase the hardy, prolific shrub, with its promise of abundant blooms and its nostalgic reminder of home "back east." From an old account book comes this item: "Purchased from Peddler—Mar. 1880—Double Lilacs, 40¢ each."

When the pioneer settlements became solidly established towns and cities, and the railroads arrived to provide swift and efficient transportation, the plant peddler disappeared. Lilacs, however, still flourish on abandoned farms and around empty cellar holes, poignant relics of the pioneer past that now give pleasure to thousands of motorists each spring.

Lilacs in the West Indies

In colonial days the West Indies seemed at least as close to the colonies as they do to the Atlantic states in this age of flight. Coastwise shipping was the quickest method of travel, and the arrangement involving West Indies sugar and New England rum was mutually profitable, and had pleasant ramifications. Oranges, lemons, and limes appeared in a surprising number of New England recipes while they were still a rare luxury in Europe. West Indian trees and flowers were introduced to our southern states. And, perhaps surprisingly, lilacs found a natural home in the Caribbean islands, and spread there rapidly.

A legend of magic virtue spread with them. When they bloomed, their flowering branches were hung above the doorways of planters' mansions and native huts alike. Their perfume was thought capable of dispelling any evil that sought to enter. The custom is still followed by some islanders, who revere the peaceful magic of nature and consider the lilac an herb of grace.

Persian Lilacs

Two thousand five hundred years ago lilacs were finding their way to Persia from Kansu province in north central China, which was crossed by two great Asian highways that led westward to the Caucasus, the Caspian Sea, and the Near East. Over these ancient trails the caravan trains carried silks, spices, musk, rhubarb roots, peach stones, seeds of all kinds, and the hardy, universally appealing lilacs. And in Persia the powerfully fragrant purple lilac, *Syringa persica*, came into its own.

Lilacs were sold in Persian bazaars, and they grew in the walled gardens that were an inseparable part of Persian homes. In such gardens, for centuries, Persians have gathered at lilac time. They spread their rugs before a single bush, whose long sprays of pale perfumed flowers light up the garden, and there they meditate, play the lute, pray, and sip sherbet. They depart refreshed by the beauty, fragrance, and mystery of nature.

The Persians gave the lilac its name, which is a corruption of the Arabian word *nilak*, the famed indigo-blue dye of Arabia, whose color is the very shade of Persian lilacs.

From cultivation in the gardens of the wealthy the lilac escaped to the hills, and grew wild in great abundance. Eventually it spread through the whole empire of Persia, which once extended southeastward to India and westward to the Mediterranean. Persian lilacs were growing on the Barbary coast in 1620 when the English fleet arrived to rout the pirates preying on British shipping. (See Gladiolus.)

John Tradescant, Sr., had accompanied the expedition to collect plants. Among those he brought back was the Persian lilac, which he later also introduced to France. But Persian lilacs do not have the robust character of common lilacs, and although they grow well enough in New England, they did not thrive in the varied and colder climate of northern Europe, where there is no record of them before 1640.

For centuries it was believed that they were native to Persia; but in 1915, after months of research and exploration, E. H. Wilson found them growing wild in Kansu province— and with such abandon that he concluded they must have originated there.

Lilacs in Europe

While the Persian lilac was finding its way from Cathay to Persia, the common lilac, *Syringa vulgaris*, reached the Caucasus and then began proliferating through the Balkans. Because it had a less dramatic flower, it was cultivated only occasionally in great formal gardens, and it eventually became a familiar wild shrub, to which the Turks gave a name translated into the English of the time as "fox taille."

The Turks, in whose gardens the flowers of the East and the West met, were the pre-eminent horticulturists of the fifteenth and sixteenth centuries. Long before the rise of the Dutch gardens, they supplied Europe with rare plants and the knowledge needed for growing them.

Around 1550 Ogier Busbecq saw lilacs blooming in the sultan's garden in Constantinople. He begged some cuttings, and sent them off to Vienna for the emperor's gardens. There Konrad von Gesner, a German naturalist traveling in Austria, was given some shoots, which he carried back to Germany.

Busbecq's old friend Mattioli also became a lilac enthusiast. The Vienna edition of his 1544 *Commentarii*, published in 1565, contained the first pictures and description of the lilac to be printed in Europe, and credits Busbecq with its introduction from Constantinople. Through Mattioli the lilac was carried to Italy and Bohemia, where it met with a warm welcome and quickly established itself.

At this auspicious stage of the lilac's career another

botanist, Charles de Lécluse, arrived at the emperor's court, as representative of the great Fugger family, merchant princes of Germany, and was later appointed supervisor of the Imperial Botanical Garden in Vienna. His position enabled him to travel widely through Turkey and the Levant, where he, too, collected rare plants, particularly lilacs, which he sent back to Vienna. Returning there himself, he established a great house, and a garden in which he cultivated every species of lilac known, and from which he sent dozens of lilac slips to the many splendid gardens developing in Germany, France, and Flanders. Lilacs spread rapidly through Europe, and in England Henry VIII boasted he had "6 lelak trees" in his garden. But they soon left the pleasure gardens of kings and noblemen, and by the end of the seventeenth century were growing in cottage gardens throughout England.

White Lilacs

The first white lilacs in England were called *Syringa candide*, according to Gerard, who notes in his *Herball* that he has both white and blue lilacs in his garden and that they increase "in infinite numbers." But the white was never prolific and is seldom referred to in old records.

In 1856 the English plant hunter Robert Fortune brought a new white lilac to London from Shanghai. This was the *S. oblata*, and it caused a flurry of interest in Europe. Half a century later, in 1904, E. H. Wilson introduced a new white lilac, *S. affinis*, to America at the Arnold Arboretum. But none of the white varieties—not even the Persians—have attained the popularity of the "old blue."

French and Canadian Lilacs

French lilacs were created by Victor Lemoine. He developed many of his hybrids from the common lilac; they are a labor of love and a revelation, bearing lavish clusters of double and semidouble flowers ten to twelve inches long and half as wide; they come in purest white, and in every shade of purple, red, pink, violet, and blue.

The Canadian lilacs, *S. prestoniae*, developed by Isabelle Preston in 1920 at the Ottawa Experimental Farm, are like a vivid footnote to Lemoine's work. They are late bloomers, extending lilac time a few more weeks, and were quickly introduced to our gardens by the nurseries of the United States and Canada.

Tree Lilacs

Japan has the giant share of the tree lilacs, *S. velutina*, although a few species are found in China and two are indigenous to Korea. The first of the species introduced to a western nation were sent to Russia by R. Maack in 1837, and became the pride of Saint Petersburg. In 1876 seeds of the tree lilac were sent to the Arnold Arboretum, and produced the first specimens successfully grown in America. Four years later, in 1880, the Jardin des Plantes in Paris acquired seeds from lilac trees in Tibet. But tree lilacs are not adapted to easy cultivation and have never become generally popular in either Europe or America, where they are confined chiefly to botanical gardens and large estates.

Another Name for Lilac

Today scientists are ignoring the old beloved name, "lilac," insisting that the correct one is "syringa." This has caused much confusion among amateurs, because syringa is familiar as the name for mock orange, of the genus *Philadelphus*.

In fact, both the lilac and the mock orange belong to the Oleaceae or olive family. The Flemish botanist Matthias de Lobel classified the lilac as syringa in 1576. The word is derived from the Greek *syrinx*, meaning "pipe," and research has shown that the folk name for lilac throughout the eastern Mediterranean is "pipe tree." Ovid, who traveled much in the

Near East in the first century, says that the pipe tree is so called because the shepherds used the hollow stems of wild lilacs to make their shepherd's pipes, and that even the pipes of Pan were created from lilac stems.

The hollow stems were also utilized by the first manufacturers of tobacco pipes in Europe, tobacco having been introduced there in 1585, almost at the same time as lilacs. The fad for pipe smoking swept Europe, and the lilac conveniently provided a ready-made stem and substantiated its old folk name.

Arnold Arboretum

Modern taxonomists may dispute their official name, but lilacs sing of spring in New England. There, appropriately, at Harvard's Arnold Arboretum, is the world's largest collection of lilacs. It includes some 500 named varieties, including 190 of common lilacs alone and some 27 hybrids. There, too, are tree lilacs, French lilacs, Persian lilacs, and white lilacs of many kinds. The man most responsible for this collection is E. H. Wilson. The Arboretum's wealth of lilacs realized his dream of Oriental beauty in New England, and lilac time there is an unforgettable phenomenon. It makes us echo his own joyous exclamation: "Lilacs have truly entered into their kingdom in the gardens of North America."

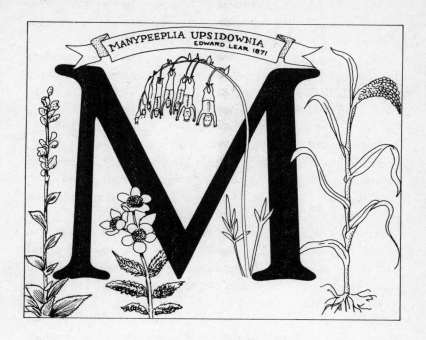

MANYPEEPLIA UPSIDOWNIA
EDWARD LEAR 1871

MARIGOLD

Centuries before the dawn of the Christian era, the holy men
of India gathered from temple gardens those marigolds that
we now call "calendulas," strung their golden glory into garlands, and hung them about the necks of their gods. The
flower was one of the most sacred of ancient India, and they
called it "herb of the sun."

From its origin in eastern Asia the herb of the sun
traveled westward to Egypt, where a practical use for it was
discovered; and this soon overshadowed its sacred history.

Saffron, a spice and a food dye that comes from the
aromatic dried stigmas of crocus (which see), was costly, and

Egyptians discovered that the herb of the sun was a most effective adulterant for saffron; it could even serve as a total substitute, and it was far cheaper to produce. Thus marigolds entered the marketplace. They were grown by the acre, and dried and sold in great quantities. They seasoned the soups and meats of the rich, were made into conserves, and were used with honey as a base for rare drinks; in the end they rivaled saffron as an Eastern spice.

In time the marigold followed the trade routes to Constantinople, the gateway to the Mediterranean and the West, and obligingly established itself in Greece. There, however, it remained a flower of the elite, a flower for temples and festivals—garlands of marigolds crowned Greek heroes and the principals and guests at marriage feasts.

As the Romans established their dominance over the world, they cultivated marigolds as widely as the Egyptians had. They used the pungent leaves and flowers in salads and preserves, for seasonings and for spicing meats, and even as a medicine. And they gave India's herb of the sun a Latin name, calling it "calendula" in recognition of its faculty, in that climate, for blooming in the calends, or first days, of every month of the year. But the early Christians, who on holy days adorned statues of the Virgin with golden calendulas, called them "Mary's gold."

They were carried by the Romans to England, where the officers and patrician rulers grew and used them as a substitute for saffron, and they quickly naturalized themselves there. Eventually, for the Saxons, they became a common seasoning in lieu of pepper and spice. The marigold also moved into monastery gardens, and its status as a dependable medicinal herb is signalized by its botanical name, *Calendula officinalis*, which meant that it was sanctioned by the office of the monastery's pharmacopoeia.

The marigold also grew in the herb gardens of medieval castles. It seasoned the knight's venison and his ale, colored and flavored his cheeses, and sometimes dyed his wool and linen. It was, as well, an ingredient in medicines, ointments, possets, and broths, and in "drinks convenient as a comforter of the heart." Calendulas were held in high repute throughout Europe, and were commonly known as pot marigolds.

New World Marigolds

By 1500 Vasco da Gama had found the route around the Cape of Good Hope, up the east coast of Africa, and across the Indian Ocean to India. The Portuguese had also established their claim to Brazil, where they found great golden-headed flowers that reminded them of the pot marigolds of Europe. They called them "marigolds," too, and on voyages in the opposite direction carried their seeds to India, the original home of the herb of the sun, where Brazilian marigolds soon flourished in temple gardens and eventually supplanted calendulas as the sacred flower of the Hindus. Today the sacred bull walks through the Benares bazaar in India with a wreath of American marigolds around his neck, and the flower stalls near the public ghats where bodies are cremated sell them by the millions. A modern commentary on global botany.

When Cortez conquered Mexico, the Spaniards discovered that marigolds were the sacred flower of the Aztecs. Even today they hold a special place in Mexican and Guatemalan culture. They decorate churches and wayside shrines, and on All Saints' Day Mexican and Guatemalan peasants strew marigold petals over graves.

The Spanish padres who accompanied the conquistadors marveled at the massive golden heads of these marigolds created by Aztec hybridists, and sent their seed to Spain in 1520, aboard the first ships returning with news of Cortez's conquest. They also sent back seeds of the dwarf red-and-yellow marigolds, which according to legend were living witness to Spain's massacre of the Indians—the red splotches symbolizing the bloodstains on the gold the conquerors had seized.

The Mexican marigolds were planted in Spanish mon-astery gardens, and proved so adaptable that their seeds were sent on to other monasteries, in France and in North Africa.

The tall golden-headed marigolds found a natural home in North Africa, escaped the walled gardens of the missions, and spread rapidly across the countryside. It is a matter of record that Charles V, who led an expedition to Algeria in 1535 to free 22,000 Christian slaves from the Moors, found the coast from Algiers to Tunis abounding in marigolds. He called them "Flos Africanus," and in honor of his victory over the Moors they were carried back to Europe, where they were known by this name until the eighteenth century.

Dwarf red-and-yellow marigolds from Mexico were growing in the royal gardens in Paris by 1598. Both these and the giant marigolds reached England a few years later, and were erroneously named "French marigolds" and "African marigolds," from the nations exporting them. English garden-ers received them with enthusiasm, and William Turner hastened to publish the first picture of the marigold in his *New Herbal* of 1551.

But much confusion was caused by these glamorous im-migrant marigolds, and much genuine resentment, too. There were many who felt that their popularity overshadowed the diligent calendula, the pot marigold that had served man so long and so well. Gerard summed it up nicely: "The French and Africane marigolds have no use but show." Nevertheless, all kinds of marigolds were popular in English gardens and herbals. They made their way into literature, too—Shake-speare often mentions them.

After two centuries of confusion, Linnaeus finally classi-fied the marigolds. He gave the Aztec species the generic name *Tagetes*, after the Greek god Tages, who first taught the Etruscans the art of divining for gold. Thus the tall golden African marigold became *Tagetes erecta,* and the small red-and-gold French marigold became *Tagetes patula,* or the jeweled *Tagetes.* The calendula, or pot marigold, simply kept the name *Calendula officinalis,* memorializing its ancient lineage and its value as a medical and edible herb.

Marigolds in Massachusetts

Under Governor John Winthrop's guidance, the Puritans brought marigolds to Massachusetts. These were of course calendulas, the pot marigolds that had been so widely and variously used in Europe that every grocer and spice seller kept barrels of their dried leaves and petals, to be sold by the pennyworth for seasoning broth, for potions, and for preserving pots of cheese and meat with an appealing color and flavor.

Pot marigolds were grown in every English colony from Maine to Georgia, and the colonists' high regard for them is manifest in recipes from treasured family files and old garden and cookery books. These are just a few:

❧ Marigold flowers and Madonna-lily bulbs may be pounded together for cure of festered felons.

❧ An infusion of marigolds in wine "soothes a cold stomache" and removes warts and moles.

❧ Marigold paste applied to the afflicted area cures toothache, jaundice, sore eyes.

❧ Marigold tea "deceives graying" and is an excellent hair tint.

❧ "Bewitching nosegays worn in the bosom of a maiden is a goodly deployment for the attraction of the male."

The gay marigold was irresistible to the disciplined Puritans. Subtly, it found its way even into their austere religion. A small bunch of marigolds was frequently carried to church; sniffed periodically, its pungent fragrance helped the ladies look wide awake during relentlessly long-winded sermons.

French and African marigolds arrived in the United

States soon after the Revolution, when great stores of flowers, ornamentals, and orchard stock were shipped from English and Dutch nurseries for the gardens of our rapidly growing country. And in 1783, African and French marigolds arrived from their native Mexico as well.

The ingenuity of the pioneer farmer led to the discovery that they had their own way of benefiting man: they were excellent insect repellents, and when they were planted among tomatoes the result was a crop virtually free from insect damage. Modern science tells us that marigolds and tomatoes are symbiotic—plants mutually benefited by growing together. (But this symbiosis, alas, is ineffectual against slugs—land molluscs as fond of marigolds as they are of tomatoes.)

Burpee's Modern Marigolds

In 1920 the famous seedsman William Atlee Burpee decided that America needed a flower that could be grown anywhere and everywhere in the nation. He chose the tall golden marigold, *T. erecta*, but there were some gardeners who objected to the pungent odor of both flowers and foliage, and a few who bemoaned the marigold's lack of variety.

Burpee collected 642 types of marigold seed, from every part of the world, but all the plants still had the familiar powerful pungency. Finally, a missionary in China sent him seeds of an odorless wild marigold. The plant proved small and scrawny, but by crossing and recrossing it with the cultivated African marigold, Burpee developed a handsome hybrid whose leaves and flowers were odorless. (To the nostrils of another tribe of gardeners, who find the characteristic aroma exhilarating, this is a deprivation. And they wonder if odorless marigolds repel insects.)

From hybridization have come the lovely shaggy chrysanthemum marigold and the flatter carnation marigold, which have won instant popularity. Odorless hybrid French marigolds as diverse as the Africans were also developed, and range from lemon to mahogany in color, with natural and incurved petals, with crests, and with flat tops.

Happily, the pot marigold has not fallen into neglect. More and more florists are forcing calendulas for winter-flowering plants, and city dwellers, who are grateful for their sunny colors and abundant bloom, have welcomed them. Furthermore, in the course of the last thirty years the hybridists have been at work on calendulas, too, and today there are 100 known varieties.

Marigolds in general have become one of our basic annuals, now grown from coast to coast, and are rivaled only by petunias and zinnias (which see). The late Senator Everett Dirksen suggested that marigolds be made our national flower, and it is not surprising that Burpee should have turned the suggestion into a campaign. We still have no national flower; but the Burpee company, undaunted, set up a standing offer of $10,000 to anyone who developed a pure-white marigold. The prize was awarded in 1975 to an amateur gardener, an Iowa widow and mother of eight, who said she was "bowled over" by her success. No wonder: the contest ran for twenty-one years, and Burpee spent $250,000 growing and testing the entries.

Military Emergency Medicine

The medicinal reputation of pot marigolds through the centuries has a firmer basis than most home remedies, and today they are listed in the *National Formulary* of medical preparations. Marigold syrup is still manufactured by some pharmaceutical houses, and calendula ointment is still prescribed for some skin irritations.

During the Civil War, when medical supplies ran short in both north and south, pot marigolds were used extensively by American surgeons to treat wounds, and proved invaluable, as Dr. W. T. Fernie testified in his *Herbal Simples*, published in 1897.

In World War I, when England was cut off from continental sources of medicine, a well-known garden writer named Gertrude Jekyll led an emergency campaign for the growing and gathering of calendulas, and sent innumerable bushels from her Sussex gardens to first-aid stations in France for use in dressing wounds. The calendula continued to serve as a hemostatic agent during World War II. Florence Ransom, in *British Herbs,* reports that great quantities of them were gathered at the time, and in some cases sent direct to hospitals.

Modern Mexican Marigolds

Mexico is cashing in on her native marigolds today. They grow by the square mile near Los Mochis, on the Pacific coast, and their bright heads are picked like cotton balls, dried, powdered, and sold to chicken-feed manufacturers. They color the egg yolks and give the flesh of broilers a rich look—essential for farmers who raise chickens in batteries and eggs on production lines. It is a mundane service, but indicative of our ingenuity in adapting the uses of marigolds to a mechanized era.

Marigolds have poured out their generous beauty to feed, and to heal, and to inspire men for thousands of years. Tomorrow's gardens, too, will surely know their magic and mystery.

MORNING-GLORY

The morning-glory has twined its way through centuries of international history and art, but Cortez may have been the first white man to observe it. We know from Bernal Díaz, who accompanied him to Mexico and later wrote a monumental account of the conquest, that it was among the Aztec seeds sent back to the monastery gardens in Spain.

With its azure flowers, heart-shaped leaves, and twining stem, the morning-glory, *Ipomoea purpurea*, seemed particularly appropriate to monastic art, and soon appeared on the pages of illuminated manuscripts and breviaries. The Age of Exploration was an age of nature in art. The perfection of violets, roses, pinks, poppies, and the heavenly blue Mexican morning-glory, and of beetles and butterflies, decorated missals and the books of hours, reminding worshipers of God's creations and the brevity of life. But the morning-glory is a prolific flower—we might almost call it a profligate one— and it escaped the confines of monastic art and monastery gardens. Soon it appeared as an accent of simplicity and contrast in the sophisticated portraits of Florentine art, and in the ornate flower paintings of the Flemish School. No more than a century after its introduction to Spain, it was climbing the trellises of many European gardens.

The Morning-glory in England

The first record of morning-glories in England is found in the garden notes of John Goodyear, set down in 1621. He describes it as "a delicate Azure with 5 straight lines inside like redd, darke crimson Velvet, a glorious shewe." His seeds, he says, come from Italy, but he does not know where they originated.

That same spring William Coys, a wealthy gentleman in Essex with a penchant for collecting foreign flowers, announced that the first tropical morning-glory was growing in his garden. He adds that the seed came from Spain by way of Guillaume Boel, a Dutchman who resided there and trafficked in "newly discovered plants." There is a suggestion of the "black market" here. As we have seen, Spain jealously guarded her American plants and any information about them. Boel also supplied morning-glory seeds for Parkinson and Lécluse.

Parkinson hastened to describe the morning-glory in his

herbal *Paradisi in Sole Paradisus Terrestris*, published in 1629, as a new flower "of faire ski-colored blew, so pleasant to behold that often it amazeth the spectator."

Its uncommon charm still held sway in eighteenth-century England. William Hanbury, the notable English garden writer, comments that the morning-glory is a "proper flower for meditation with its early budding, noon blooming and certain wilt by evening." He adds that it is often called "Life of Man" through rural England. And so it is, here and there, to this day.

Medicinal Morning-glories

Another Mexican morning-glory welcomed in Europe was the *Ipomoea purga*, an invaluable medicinal purgative, which the Indians used long before the Spanish came. It was in wide use in Europe by 1735, and was later imported to the United States.

William Houston, who went to Mexico for the London Apothecary Company in 1733, sent back seed of the *I. jalapa*, another, milder, but nevertheless important Indian medicinal of the morning-glory family. It was imported to this country from Europe at a later date.

An Unexpected Morning-glory

Yet another morning-glory is the popular wood rose, which was introduced to the Hawaiian Islands from Ceylon. It was probably taken there by the laborers from India who went to the islands in the nineteenth century to work in the sugar fields. Like other morning-glories, it soon spread rampantly over bushes, fences, roadside trees, and shrubs.

It was originally called "Ceylon morning-glory" and "Spanish arbor vine"; "wood rose" was adopted as a trade name when its peculiar beauty made it profitable commercially. Today it is cultivated in quantity and exported to mainland florists.

The wood rose is a perennial. Its large, seven-lobed leaves, sometimes measuring eight inches across, suggest horse-chestnut leaves and are quite unlike those of the annual morning-glory. In autumn, the eight- to twelve-foot vine bears an abundance of small, drab, yellow flowers, often opposite the leaves. When the true petals drop off, the calyx, which covers the large round seed pod, begins to develop, until the whole resembles a large, cream-colored button. When it dries, the calyx opens and the "flower" looks like a delicately carved, highly polished wooden rose. The whole development takes from twelve to sixteen weeks; the beauty of the mature seed pod will last for many months.

Although wood roses are found growing wild in tropical regions around the world, Hawaii is the only one of the fifty states into which they have been successfully introduced.

The Morning-glory in the United States

The morning-glory was brought to the American colonies about 1700. It responded happily to colonizing, and in the next century grew in many gardens from Salem to Charleston.

The first mention of morning-glories in American garden literature is in Bernard McMahon's *The American Gardener's Calendar*, published in 1806, where it is called by its botanical name, *Convolvulus Ipomoea purpurea*. McMahon, a well-known Philadelphia nurseryman who specialized in collecting and selling new and rare flower seeds, was the first to advertise nasturtiums (which see) as well as morning-glories.

By 1850 this *Convolvulus* had gained the familiar name "morning-glory" from its habit of opening with the rising sun. As morning-glory it was sold by flower peddlers through the countryside, and also by small-town milliners, who served a dual role, with windows offering both flower seeds and spring bonnets.

One old flower book complains that seedsmen advertised

the morning-glory with less judgment than experience, because they had proved to spread below ground as well as above.

In America, the abandoned character of the tropical morning-glory was emphasized by the invasion of its European relatives, the bindweeds and the dodders. The best known of the bindweeds, the hedge bindweed, *Convolvulus sepium*, often called the "wild morning-glory," and the lesser bindweed, *C. arvensis*, arrived in the colonies as stowaways in imported grain and garden seeds, and spread quickly in the virgin land. Flax dodder, the *Cuscuta epilinum*, sprang up in the settlers' first flax fields. Today perhaps a dozen Old World dodders wind their parasitic stems around the wildflowers and shrubs of America's fields and roadsides.

With such embarrassing relations and the morning-glory's own aggressive response to cultivation, it soon lost its popularity as a garden flower in America. But modern nurseries have rescued it, and hybridists have disciplined its opulence to produce less rampant vines and larger flowers in an array of colors from heavenly blue, crimson, and carmines, to bright pinks and whites.

Another *Ipomoea* from America's southlands is the moonvine, *I. noctiflora*. Its large pure-white flowers, delicately scented, open in the evening and close in the morning—a reverse morning-glory. Hybridists have also developed this annual moonflower in shades of palest lilacs and subtle pinks to add mystery and magic to your garden.

MULLEIN

Our roadsides are like the borders of church linens, embroidered with elaborate detail and changing their colors with the seasons. Dominating the background of the design along many railroad tracks and highways are the tall, gray-green clumps of mullein. The elegance of the broad velvet leaves seems in strange contrast to the ragged flower head that rises gauntly above them. But research reveals that this unkempt appearance is part of an orderly design in the mullein's complex sex life.

Its terminal spikes are crowded with buds, which open irregularly throughout the summer. To increase the careless appearance, each one flowers in two stages. The three bottom petals open first, forcing the pistil to bend forward as they unfold. When a bee lights on the edge of the petals, the projecting pistil scrapes from his back the pollen that he has gathered from the stamens of another flower. Then, laden with pollen, the pistil droops. It has been fertilized. Its work is done. At this instant the two top petals spring open, revealing four hairy stamens laden with pollen. The bee, seeking honey at their base, shakes the pollen from the stamen. It falls on his furry back and he carries it off to the next flower, thus ensuring efficient cross-pollination. What marvelous, detailed, perfectly timed mechanisms so that one weed may reproduce and thrive! Natural magic, indeed.

The sexuality of plants, and ways in which their structure ensures abundant seeds, was first demonstrated in 1694 by Rudolf Jacob Camerarius, a German botanist and physician, and a professor at the University of Tübingen, where he was director of the botanic gardens. It was a revolutionary idea, and it helped establish botany as an independent science in the next century.

History

Mullein has followed man across the continents. It was known in Asia, Africa, and early Europe before recorded history, and has been used through the ages for a variety of purposes. Its official name, *Verbascum thapsus*, is a corruption of *barbatus*, an old Latin word meaning "bearded" and refer-

ring to the hairy stamen, and Thapsos, a Greek island where
in ancient times mullein was gathered in great abundance.

The Greeks made common use of the mullein's woolly
leaves for lamp wicks. The Romans, elaborating, turned the
tall strong stems, topped by seeded spikes, into natural
torches. Dipped in tallow, they burned long and efficiently,
and were used for public lighting, particularly in funeral processions.

The Roman legions undoubtedly carried mullein
through Europe and into Britain; the Anglo-Saxons used it
widely not only for light but also as a medical herb. A tea
made from boiling the woolly leaves was used for centuries
for pulmonary diseases, particularly in farm animals, which
gave the plant its rural name, "bullock's lungwort"; and an
infusion of mullein flowers was used for catarrh.

During the Middle Ages the power to control demons
was imputed to mullein—perhaps because of its ghostly appearance in moonlight, when its great gray-white leaves,
rustling in the wind, take on an unearthly quality.

As Christianity was established in Europe the church
forbade cultivation of plants with suspect powers, and collected all available medicinal herbs in the monastery gardens.
Mullein was grown in such gardens throughout Europe, and
was put to many practical uses, dispensing its comfort and
healing for a thousand years. Then, as the monasteries began
to break up in the sixteenth century, mullein moved into cottage gardens and was used as a remedy for many ills: as a
poultice for toothache and neuralgia, as a means of relief for
stomach cramps and gout, as a medication for farm animals.
Its large woolly leaves were put to a quite different practical
use as well. They reinforced thin boot soles in winter and
were worn inside shoes to relieve chafed feet. Thus mullein
gained the common name "feltwort."

The Elizabethan herbalist Gerard grew mullein in his
garden, and gave it much space and a block print illustration
in his *Herball*. He says "mullein" is a corruption of the old

name "muleyn" or "woolen" and that this useful plant grows much along London highways "as about the Queen's house at Eltham." It served in various parts of England as a hedge plant, and was commonly called "ag," a corruption of the Anglo-Saxon *haege*, a taper or torch.

Mullein in America

Mullein seeds were among those brought by the Puritans to Massachusetts and planted in the first New England gardens. Perhaps Governor Winthrop was influenced in his selection of essential herbs by an ardent Puritan radical named Nicholas Culpeper, whose herbal *The English Physitian Enlarged* describes the good and bad qualities of herbs and how to prepare them for home cures. The publication of this book in 1649 enraged London medical men, but its usefulness and popularity are established by the fact that it was reprinted in 1947.

It was invaluable to the Puritans, who had to rely primarily on their herb gardens for medication. It prescribed mullein for the relief of diarrhea, rheumatism, and rough warts. Throughout the colonial period, mullein was confidently used for lung and throat disorders, and its chopped leaves were employed in hot poultices, a practice from which it derived such names as "flannel wort" and "velvet plant."

John Josselyn, in a description of a Salem garden that he visited in the late 1630's, lists parsley, sage, thyme, sorrel, spearmint, fennel, mullein, and other medical herbs as growing on either side of the garden path. On the later of his two visits to New England he noted that many such herbs had already escaped to the fields and roadsides, among them chickweed, nettles, plantain, and mullein.

John Randolph lists mullein in his *Treatise on Gardening* as one of the essential garden herbs, along with cress, camomile, feverfew, garlic, onions, and Roman and Egyptian lettuce, which would thrive in Virginia.

Between the Revolution and 1800, 500,000 pioneers were

moving from Virginia and the seaboard colonies across the Alleghenies to the Ohio valley, carrying their herbs with them. Most of the earliest homesites have disappeared, but the western movement can be traced by the mullein and other immigrant weeds that still come up in the grass where cabins once stood as settlers pushed across the mountains, the Mississippi, and the Great Plains.

Today, in rural areas of the Appalachians, a soothing ointment is still made of tallow and mullein, and a mucilage is produced by boiling down its leaves. In the Ozark Mountains it is commonly used as a rural cure for asthma; the leaves, gathered in July and August and dried, are smoked with apparent pleasure. The juice of the mullein is also dropped into the ear to relieve earache and deafness.

As one drives farther west mullein remains a familiar roadside companion. It thrives in the foothills of the Rockies; it runs along the railroad tracks; it flourishes on sunny hillsides of abandoned mining camps and homesites. Thousands of clumps of gray-green mullein stand stalwart in the Grand Canyon, tossing their unkempt heads and scattering millions of seeds to assert their place in one of our noblest scenic wonders.

Anatomy of Mullein

Mullein is a biennial. It belongs to the Scrophulariaceae or figwort family, a group of plants that were thought to cure the dread scrofula of the Middle Ages. There are a hundred species in this family, including the beautiful moth mullein and digitalis. (See Butter-and-eggs; Foxglove.)

In its first year the mullein presents no more than a broad rosette of woolly gray-green leaves; the second year it sends up a strong straight stem two to six feet tall, and sometimes branched. The upper leaves are smaller than the base leaves; they are pointed, and, peculiarly, they extend down into the base of the leaf below.

The long terminal flower head is crowded with the buds that open by degrees throughout the season. The flowers are small, round, five-petaled, and bright sulphur yellow. The calyx is woolly. The numerous seeds—which are enclosed in a round, woolly, two-celled capsule—are extremely interesting in design, with wavy ridges alternating with deep grooves.

The mullein, like most figworts, has a strong taproot to anchor it.

Future of a Sumptuous Alien

The candelabra mullein, *V. olypicum*, is still grown in some English flower gardens, but nurserymen have lost patience with it because plants grown from its seeds never come true to form. In America no mullein has made a domestic place for itself since the colonial herb gardens.

Although it was introduced as a medicinal herb, it has never been recognized by the United States Pharmacopoeia, and it has long since been removed from England's. But today there is a resurgence of interest in herbs on the part of pharmaceutical companies, and a growing emphasis on research into ancient cures. This intrepid alien weed has invaded every state in the union and occupies thousands of American acres, from which treasures still undiscovered may pour in the future. Already one wholly new use has been found for mullein. Its silvery mats are now planted in gardens for the blind, not for visual beauty or for fragrance, but for the pleasure the velvet leaves convey to the sense of touch.

A weed is a plant whose virtue is forgotten.

NASTURTIUM

The conquistadors and the priests who accompanied them must have been the first white men to behold nasturtiums, hanging from vines in the jungles of Mexico and Peru. And nasturtiums were among the first floral immigrants from the New World to Europe.

Nicholas Monardes, a doctor of note who lived in Seville, became intensely interested in the plants that were arriving from the New World, and for several decades collected all available material about them. He had access to the official reports of the explorers; he personally interviewed returning sailors, traders, and priests; and he begged and bought seeds and plants, with which he started an experimental garden.

Among the first seeds he sowed was the odd one of the nasturtium, which looked rather like a caper. In 1569 he published all his material in the first herbal ever written about plants of the New World. It included a description of the nasturtium, and the first drawing of it to appear in print. Interest in America and its vegetation was intense and widespread, and Dr. Monardes's herbal was soon translated into French and English, and sold widely in both countries. Its delightful English title was *Joyfull Newes Out of the Newe Founde Worlde*.

Apparently the first nasturtiums to bloom in Monardes's garden were red, because he calls them "flowers of blood." They attracted much attention, and soon the seed was being planted in many Spanish gardens. It traveled quickly to Portugal, and then to the royal gardens of France. And in 1577 Gerard wrote that the gardener to the king of France had sent him some nasturtium seed. It is difficult to imagine, in these days of instant communication and swift transportation, the excitement this jewellike flower from the jungles of South America caused among English gardeners of the sixteenth and seventeenth centuries.

Illustrators, too, have found it irresistible. Its twining vines, exquisite flowers, and round leaves made it a perfect subject for old herbals and the illuminated pages of psalters, gospels, and books of hours. Perhaps no other plant was used for this purpose so often.

When the nasturtium arrived in England it lost its Spanish name, "flower of blood," and was generally called "Indian cress." But soon it acquired, from rural gardeners, other names—"canary flower," "yellow larkspur," and "lark's heel." Its long-spurred corolla, which suggested the larkspur, clearly gave rise to the two latter, but its pungent odor won it a name borrowed from the water cresses—"nasturtium," which means "nose-twister." The new arrival's quick popularity demanded that it be classified and given an official name, too. Because water cress had already pre-empted *Nasturtium officinale*, the garden flower became *Tropaeolum majus*. The first part derived from the Greek word meaning "to twine," and *majus* is added because it is the larger of the garden nasturtiums. There is a dwarf variety, native to Colombia, which is called *Tropaeolum minus*. But the popular names prevail—"nasturtium" for the garden flower, "water cress" for the aquatic, pungent salad green.

The nasturtium is a distinctive botanical specimen, and has always refused to be easily classified. Asa Gray says our garden nasturtium belongs to a group of intermediates between the balsam and the geranium.

Nasturtium as Food

Part of the nasturtium's popularity in Europe can be attributed to its food value. It belongs to the same family as several land cresses, and as Indian cress it was widely used for salads. Its leaves, stems, and flower buds are crisp, tangy, and delicious. Many references to the delights of nasturtiums as food are found in old books. John Evelyn discourses at length on the joys of nasturtium salad.

Water cress, *N. officinale*, is a member of the mustard family, and an unrelated species. (Botany can be confusing.) It was used as a specific against scurvy, and the directors of shipping companies (botanically confused, perhaps) believed that garden nasturtiums might serve the same purpose. So nasturtium seeds were pickled by the barrel (imagine how many there would be in a barrelful!) and carried aboard sailing ships. They appear to have been at least as effective as *N. officinale*, because they were used as an antiscorbutic for many decades.

Europeans were not the first to use nasturtiums for food. In tropical America the Indians ate the leaves, stems, and flower buds as a delicacy; they also ate the tubers of a variety called *Tropaeolum tuberosum*, which were dug and stored for winter eating by the Peruvians, just as potatoes were.

Pickled nasturtium seeds are still a popular condiment in Europe, and used as a substitute for capers, and many old recipes are extant. Here is one from Mrs. Beeton's *Book of Household Management:*

PICKLED NASTURTIUMS
(A very good substitute for capers)

Ingredients.—To each pint of vinegar, 1 oz. of salt, 6 peppercorns, nasturtiums. Gather the nasturtium pods on a dry day, and wipe them clean with a cloth; put them in a dry glass bottle, with vinegar, salt and pepper in the above proportions. If you cannot find enough ripe to fill a bottle, cork up what you have got until more are fit: they may be added from day to day. Bung up the bottle, and seal or resin the top. They will be fit for use in 10 or 12 months; and the best way is to make them one season for the next.

Seasonáble.—Look for nasturtium-pods from the end of July to the end of August.

The Nasturtium in America

The excitement about nasturtiums in the seventeenth century coincided with the departure of colonists for North America; so the jewels that the padres had found climbing over the rocks of the Andes were soon climbing on the stone walls of Virginia. The exact date of their arrival is not recorded, but we learn from John Randolph's *Treatise on Gardening* that they were well established in Virginia before Williamsburg was founded. He tells us that the nasturtium was one of the first flowers to be planted in the garden of the Governor's Palace, in 1698, and that it was still valued as a rarity, worthy of cultivation both for its beautiful flowers and for its excellence in salads.

To McMahon's Philadelphia nursery goes the distinction of having first listed nasturtiums in a seed catalogue. That was in 1806; McMahon recommended them highly, and they caught on almost at once. They were easy to cultivate in many types of soil, their round-leaved vines were decorative, and their colorful bright flowers bloomed a month after planting and continued until frost. They became one of the flowers of the pioneer era, spread quickly across the plains, and were eventually grown from Maine to California.

They became so familiar, however, that their popularity waned. Then in 1840 an English plant hunter named William Lobb discovered a new nasturtium, *Tropaeolum azureum*, in Chile. Five years later he found another, *T. speciosum*, in Patagonia. This spurred new experiments in hybridization, and led to the development of the Tom Thumb variety, to which gardeners both here and abroad took an instant fancy.

In 1928, J. C. Bolgier, a California nurseryman, discovered what he called the "Golden Gleam" nasturtium. He began its propagation at once, and in 1931 it won an international prize at the Chelsea Garden Club in London. One year later his nursery sold three tons of Golden Gleam seed.

Nasturtiums climaxed their long, brilliant reign when David Burpee developed the double multicolored nasturtiums and introduced them to American gardeners. This elusive form had originally been noted in 1724 by Phillip Miller, but it seems to have been considered a sport. There is no further record of it until 1756, when John Hill wrote in *The British Herbal* that double nasturtiums appeared occasionally and were esteemed a great rarity. Two centuries later they bloom in gardens across the nation.

Seed houses now focus on the new hybrid nasturtiums, bred to smaller leaves on shorter stems and to an abundance of many-colored, perfect flowers. Climbing varieties are still available, but even these are very different from the old-fashioned ones, which bore a great many leaves and relatively few flowers, often irregular in shape, and were limited to yellows and reds. Nasturtiums are now staging a real comeback, and they may soon become as widely popular and as intensely appreciated as when they were first introduced to Europe and America.

OLIVE

I gazed with wonder at the display on my favorite fruit stand. It featured boxes of ripe olives, totally unprocessed, which could be bought by the pound and prepared as the buyer wished. Even the suppliers of olives had bowed to the vogue for natural foods. Bright folders detailing recipes and steps in preparation were at hand but, as we shall see, the unsuspecting buyer was letting himself in for a great deal of work.

Unlike most raw fruits, tree-ripened olives are too bitter to eat. They require such painstaking preparation that one wonders how man discovered they were edible at all. But in fact they are one of his oldest sources of food and oil, and throughout the ancient Mediterranean world they were a symbol of peace and plenty.

History

The olive is a small evergreen, native to Asia Minor, and commercially the most important member of a varied family of trees and shrubs, the Oleaceae, which includes lilacs, privet, and forsythia. This true olive, *Olea europaea*, is frequently mentioned in the folklore and myths of the Mediterranean, and in the oldest Bible stories: it was an olive twig that the dove brought back to the ark to assure Noah that the flood was over and that a green world awaited him. In Greek myth Athena, goddess of peace as well as war, of the arts and crafts, and ultimately of wisdom, gave the sacred olive tree to mankind, and it became a symbol of honor throughout Greece and the Roman Empire. Heroes were crowned with olive wreaths, Olympic victors wore olive leaves, and an olive branch became a symbol of peace, a flag of truce, through untold centuries.

Rome carried the olive to Spain and Portugal, which, like Italy, developed their own varieties. Rome also introduced the olive to northern Europe and England, but it demands a hot dry climate and its northern limit is the southern Crimea. Here the descendants of olive trees planted centuries ago by Roman and Greek settlers still thrive.

Throughout the Middle Ages wealthy English nobles imported olives for their feasts—to be eaten after the meal, so their peculiar flavor might neutralize the aftertaste of food, permitting the host and his guests to savor their wine to the fullest.

The oil of the olive has been a precious product for untold ages. Invaluable where animal oils were scarce, it has been used as a food, condiment, and pharmaceutical, and was long a measure of man's wealth. The oil is pressed from both the pit and flesh of uncured olives, and the yield varies with the method and the condition of the fruit. Some varieties of olive have been developed especially for their oil content, and olive oil is used extensively in lubricants and soaps today, and remains a kitchen staple.

Olives in America

Spain brought olives to America. They were among the first fruits planted in the West Indies, to provide oil for the colonists, and the dry warm climate there and elsewhere in Spanish America proved perfect for them. Olive trees had flourished in hacienda orchards and mission gardens for two centuries when the padres established the mission trails through California and made the first planting recorded in North America at the mission of San Diego de Alcola, about 1769.

Cuttings of these trees were carried to other missions, and began the successful propagation of California olives, which, traditionally, have been called "mission olives." But olives did not become popular in the United States until the turn of the century, because primitive curing methods limited their commercial distribution.

About 1900 the University of California at Berkeley began a program of research into the curing of olives. The result is a modern, efficient variety of methods that have established the importance of the olive industry in the United States and created a year-round demand for its products.

Today, most American-grown olives are processed as so-called ripe olives, either black or green in color, and packed in cans. But in fact they are not literally ripe; they are picked green from the tree, as are almost all olives, anywhere in the world, that are grown for eating purposes. The color depends on the processing.

Ripe black olives and ripe green olives alike are picked in October, and may be held in brine for several months thereafter. When the processing resumes, they are soaked in a dilute lye solution, partly to get rid of the bitter taste, from two to four days. During the lye treatment, those that wind up black are treated vigorously and continuously with jets of air, a technique that oxidizes the skin and meat to its dark color. Those that are to be marketed as ripe green olives are not aerated, but are kept immersed in the lye solution to preclude oxidation. In either case the final steps are the same: the olives are washed thoroughly to remove all traces of lye, and then packed in a dilute salt solution in cans, which are heated for about one hour at 250° F., cooled, and sent to warehouses.

Olives marketed in glass jars or bottles as Spanish green olives have undergone a wholly different process, essentially one of fermentation through the use of bacteria that form lactic acid. Following harvest they are graded for size uniformity, and placed in barrels or vats with a brine solution to which a little sugar has been added for the bacteria. A particular strain of such bacteria may be introduced artificially, by inoculation, or the vats may be allowed to receive wild or natural bacteria from the air; the medium is selective, and only the acid-forming bacteria will grow in it. The fermentation process, which is conducted at a temperature of 100° F. or more, continues for several days or even weeks, until the correct amount of lactic acid is formed. The olives are then removed, and packed in jars, as they are or after pitting, stuffing, spicing, or other final steps.

The one exception to the processes described above is the so-called Greek or Italian olive. Olives to be cured this way are not picked until they are tree-ripened, by which time they are dark purple, soft, and juicy. They are then put down in rock salt and left for several months before being packed in olive oil, marketed, and eaten by those who scorn the black "ripe" olives from California. We must assume that the natural-food devotees who buy those raw, ripe olives prepare them in this fashion, and wait patiently to relish the result.

The green Spanish-style olives account for about 80% of the present U.S. market; the remaining 20% is shared more or less equally by the American ripe olives, black and green. "Spanish" olives are produced in limited quantities in California, but the current demand for olives of any kind does not really exhaust the supply. Leaders of the industry, however, foresee a bright future. United States technology is in the vanguard, and acreage planted to olives is increasing. The time may come when Americans will consume not only American ripe olives that were picked unripe from the tree, but Spanish olives grown 6,000 miles from the Iberian Peninsula.

ORANGE

Oranges were brought to America by Columbus in 1493—the first of the fruit trees introduced by Europeans that have been successfully and continuously cultivated ever since. His handful of seeds has burgeoned into 50 million orange trees in the United States today. Millions of others grow on the Caribbean islands and in Central and South America.

It was no coincidence that Columbus chose the orange for his initial attempt to introduce Old World fruit to the West Indies. Behind that choice lay strong economic pressure, national rivalry, and financial support for his second voyage.

Why Columbus Planted Oranges

Oranges began their romantic history in the remote mountain valleys of Indochina perhaps 3,000 years before Christ was born. Throughout antiquity their golden color marked them for the court and the temple, and made them generally coveted.

Eventually orange seeds were carried by caravans on the old routes from the Orient to the Persian Empire and the eastern Mediterranean. But they remained rare and costly, and tales about them became entangled with the legends of Greece and Rome. Legend has it, for instance, that the golden apple the goddess of fertility gave to Hera when she married Zeus was an orange, and that Hera planted its seeds in the Garden of the Hesperides. Rare and sacred, used in fertility rites through the ages, the orange was clearly the fruit of the gods.

When the Moors invaded Spain in the tenth century, they brought oranges with them—but as sacred fruit, to be used only in religious rites, for medicinal purposes, or as an exquisite flavoring in food and drink. They, too, incorporated them in fertility rites, crowning their brides with orange blossoms, as we do to this day. They grew them only in their walled gardens, guarding them so jealously that any Christian who ate or even touched an orange did so on pain of death. These orange groves were perhaps the most valuable legacy the Moors left Spain when they were driven out after 500 years. During those centuries, oranges and sugar had become the most valued trade commodities in Europe; Spain watched Portugal and Venice grow rich on them while she fought the Moors.

It was at this moment in history, in the very year Ferdinand and Isabella finally vanquished the Moors, that Columbus returned from his first voyage to America. He brought from his bankrupt venture no gold, no spice, no jewels, no treaties with Eastern potentates; only one battered ship, a starving crew—and an idea.

By the time his disappointed sovereigns reluctantly granted him an audience, he had sized up the situation in Spain. He did not dwell on the gold and spices he had failed to bring back. Instead he offered them the green gold of plants, the tropical climate of the West Indies, unlimited space for sugar and orange plantations, islands far richer than Portugal's envied acres in the Azores.

The king and queen listened. There was a grandeur about the pristine world Columbus described that captivated their minds. It was perhaps the doorway to new empires for Spain.

Columbus had not expected to discover a new world, but the verdure of the West Indies had instantaneously enchanted him. The records of his first voyage show he often wished he had taken with him a botanist to identify and evaluate the plants of the new-found land, instead of the Hebrew interpreter whom he had engaged to translate treaties with Oriental potentates.

His first voyage had been financed in the hope that it would bring Spain a spice monopoly in the East. His second was financed in the expectation of establishing sugar and orange plantations in the Caribbean islands. Unwittingly he became America's first explorer and importer of plants. From the time of his second voyage, plants—not gold or spice or titles or honors—became his all-absorbing interest. Plants dominate his reports and letters.

When he sailed to America in 1493 his flotilla of seventeen ships stopped at Gomera, westernmost of the Canary Islands, where he purchased wheat, rye, peas, and grapevines —plants essential for feeding a permanent colony. He also purchased orange seeds and cuttings of sugar cane—plants essential to Spain's commercial competition.

He laid out his plantation in Hispaniola—the island now divided between Haiti and the Dominican Republic—with sugar cane, rows of orange seeds, and native pineapples (which see). His plantings, he reported to Isabella, throve mightily. "Thus," wrote Bartolomé de Las Casas, the sixteenth-century missionary and historian, in his *Historia General de las Indias,* "was first introduced to the islands of the West the fruit of the Hesperides." A new chapter in the history of the orange had begun.

Oranges in the Era of Exploration

The success of Columbus's experimental plantings inspired Spain with the idea of dominating the European orange market. The Council of the Indies immediately ordered that every sailor voyaging to the New World carry with him as part of his equipment one hundred orange seeds, to be planted around any harbor where Spanish ships might lie. Portugal, Arabia, and the Netherlands watched this move with interest, and soon ordered their crews on ships sailing to the Orient to plant orange seeds wherever they dropped anchor on the long journey around Africa. This would not only establish orange groves conveniently located for shipping purposes, but would provide the crews with the fresh fruit that helped to control scurvy.

Vasco da Gama sailed under the "orange rule" in 1497, on the first successful voyage to India around the Cape of Good Hope. His sailors planted orange seeds at every landfall, for use by future expeditions.

Hernando de Soto carried orange seeds on his explorations from Tampa to the Mississippi, and planted them at likely camping spots.

Captain Cook, on the first voyage to Australia, carried orange seeds and planted them on various islands and around harbors in the Pacific where his ship anchored. The seeds had been gathered in Rio de Janeiro, from trees that Portuguese sailors had planted, by the naturalist Joseph Banks, who accompanied Cook on his stupendous voyage. (See Eucalyptus.)

George Vancouver, who sailed on Cook's second and third voyages, planted the first orange seeds in Hawaii in 1778, on the island of Kauai. They flourished there, and large shipments of oranges were sent to California before its own orange groves were widely established.

Through the impetus of sailing masters in the era of exploration, oranges were planted in the subtropics around the world, and played a lively part in history.

Sweet Oranges

All the oranges planted by the early explorers were bitter oranges, prized for flavoring, preserves, and medicinals. No one in Europe had even tasted a sweet orange before 1616, when the first ship of the Dutch East India Company reached China. There the voyagers discovered the sweet orange, found it delicious, and collected some seeds; but, fearing they would not stay viable on the year-long trip home, carried them to Goa, the enclave the Portuguese had wrested from the sultanate of Bijapur in 1510, whose harbor was the only one on the west coast of India open to Dutch ships. The orange seeds were planted there; later the young trees were transported to Portugal and carefully established.

"These sweet oranges," says E. H. Wilson in *China, the Mother of Gardens,* "were the first Chinese plants of record taken to Europe."

The fruit of the sweet orange trees received such acclaim that both the Dutch and English East India companies pursued the commercial possibilities of importing other plants for the gardens of Europe. Soon every East Indiaman carried her quota of flowers, trees, and shrubs from the Orient. Thus oranges had opened up a new source of garden wealth. It remained the monopoly of ship captains until the end of the eighteenth century, when professional plant hunters were sent to far places by national governments, nurseries, and botanical gardens.

Oranges and European History

Not long after its advent, the sweet orange began playing a large role in the history and economics of Europe. By 1654 the Dutch had thriving citrus plantations in Africa; Portugal had planted large groves in the Azores and Madeiras; and Spain was cultivating oranges in Andalusia and Valencia as well as in her American colonies.

One key to the romance of the orange has always been its color. There are many fruits that are yellow, or yellow blended with red, but the orange is unique in being entirely golden. The word "orange" itself has incorporated into it a corruption of the Latin word for gold, *aurum,* although it derives originally from the Persian *nārang.* This became the Arabic *nāranj,* the Spanish *naranja,* and the Provençal *auranja,* into which spelling the Provençal *aur,* for gold, insinuated itself. Centuries later, when the city of Auranja in Provence became a center of orange growing in southern Europe, the name of the city and the fruit became synonymous.

The Dutch statesman known as William the Silent inherited the principality of Auranja in 1544. In 1677 a descendant of his, another William of Auranja, married an English princess named Mary. She was the Protestant daughter of the Catholic Duke of York, who became King James II. William joined forces with James's Protestant opponents, and in the end supplanted his father-in-law on the British throne. His name was anglicized to William of Orange, and his supporters were known as Orangemen.

Sweet oranges, by a historical coincidence that seems almost too pat to be true, were at this precise moment being introduced to England, and became a status symbol among the wealthy and the nobles. Because they would not grow in England, orangeries or large conservatories were added to many castles and manor houses, and in them orange trees were grown in tubs. (Only recently was the ancient orangery at Windsor Castle converted to an indoor swimming pool by Queen Elizabeth II and Prince Philip.)

To avoid the religious turmoil of the eighteenth century, thousands of British and North Irish migrated to the colonies. A great number of Protestant settlements sprang up in North and South America and in South Africa flaunting the name Orange and the Protestant faith. Among them were Orange, New Jersey, and Fort Orange, New York, which later became Albany; and dozens of Orangevilles, Orangetowns, Orange Rivers, and Orange Creeks found a permanent place on maps.

Oranges in America

The sweet orange was well established in Spanish America by the early 1600's, and soon crowded out the bitter oranges. In many orchards bitter-orange trees were cut down and sweet oranges grafted onto their roots, so that hundreds of groves of sweet oranges were soon bearing abundantly.

Eventually the supply outran the demand. Many groves were abandoned and many trees went wild, as they have a strong tendency to do; and the Indians seized the opportunity to plant oranges about their camps. Today the ancient groves that still thrive along lakes and rivers of the south provide the best clue to the location of old Indian settlements.

In 1840 the first Spanish ambassador to the Republic of Mexico and his American wife kept a diary of their travels through that lovely country. Again and again they noted extensive old orange groves abandoned, stately old trees gone wild and loaded with waxy fragrant flowers, or the ground covered with golden fruit. Wild oranges are still as common in Mexico as wild apples in New England.

Hernando de Soto had claimed Florida for Spain in 1539 by planting oranges about his camps. Later the Spanish missions had maintained thriving groves, the most famous of which were perhaps those of Saint Augustine, which were laid out in 1565. But Florida remained sparsely settled, and oranges did not become a commercial venture until the United States took over the colony in 1820. From then on the cultivation of oranges expanded rapidly, particularly along the Saint John and Indian rivers. It is interesting to note that among these relatively early orange growers was Harriet Beecher Stowe, the author of *Uncle Tom's Cabin*, who bought a plantation after the Civil War to help the former slaves—a venture that proved unsuccessful.

Oranges went west with the southern pioneers. Groves of sweet oranges soon flourished around the Gulf of Mexico, particularly in Louisiana, and by the late 1860's they began appearing more often in northern markets. At first they were bought as a rare Christmas treat, an exotic dessert, a fresh fruit for a special occasion in the long winter. But the demand increased, new methods of production and transportation developed, and the sweet orange became a Florida success story.

Navel Oranges

When Father Junípero Serra and his Franciscan monks began establishing a chain of missions in southern California in 1769, they carried orange seeds with them. By the time the United States took over the area in 1848, the beautiful valleys were fragrant with hundreds of groves of sweet oranges. These California crops, ripening late, complemented Florida's early spring and summer oranges and helped meet a growing demand for the fruit.

In 1870 our young and ambitious Department of Agriculture learned that a new type of orange had just been discovered in Bahia, Brazil. It was seedless, and bore a distinct "navel" on the blossom end, beneath which there often snuggled a sort of baby orange. It had originated as a bud mutation of wild seedlings, which were offspring of old sweet-orange trees imported by early Portuguese. The Department of Agriculture bought twelve of these seedlings and planted them in various orange-growing areas, but only one of them prospered. This tree, growing in Riverdale, California, in due course bore fruit—bright golden navel oranges, sweet and seedless. Grafts from it provided thousands of young trees for orchard stock, and established early groves of navel oranges in California and Arizona. The matriarch tree itself was still thriving in the twentieth century.

The navel orange, providing a fall and winter crop, filled a gap in the orange market, and oranges have become a year-round fresh fruit, available every month in the year.

Science and the Orange

Today orange growing is a major aspect of our agriculture, backed by research and encouraged by an expanding market.

But the successful grower has discovered that he must digest the vital statistics of the orange to compete.

The orange will grow impartially in high dry places or in low alluvial ones, but it needs fertile soil. Temperatures below 60° F. arrest its growth, but its thick oily skin adapts it to subtropical climates. Botanically speaking the orange is a berry, a simple, pulpy fruit with eight to ten seeded sections. The tree is an evergreen, low branched, with large elliptical leaves; uninhibited, it grows tall and stately, but those in our orange groves have been disciplined for efficient picking, so that a grove has a uniform height of around thirty feet. A tree will begin bearing early in its second or third summer, and will produce a commercial crop in six to eight years. Some trees are reputed to have survived for three centuries.

There are many classes of oranges, but they all originated in the Orient and are members of the citrus branch of the family Rutaceae, or rue. Today there are seventy varieties of sweet oranges alone. The sweet orange is classified as *Citrus sinensis;* the sour orange as *C. aurantium;* the bitter orange as *C. amara;* and the bergamot orange, which produces bergamot oil, as *C. nobilis.* Among the *C. nobilis* oranges are the king oranges, mandarins, and tangerines.

Growers have found that the color factor poses the greatest problem in orange production. In order to get the fruit to distant markets before it spoils it must be picked green though fully mature. But oranges do not ripen further on the way to market, as many fruits do, and nobody will buy a green orange. The rare gold color appears naturally only after the orange has ripened on the tree and been exposed to a period of coolness. Until that occurs the green chlorophyll in the skin predominates, but a drop in temperature releases the carotene and turns the fruit a bright golden color, just as the leaves of northern trees turn to bright reds and yellows as fall grows chilly.

Early commercial growers discovered that green oranges

stored with McIntosh apples ripened quickly: one "mac" gave off enough ethylene gas to ripen a dozen oranges in cold storage! Apple "degreening" was not efficient enough for modern economy, but nature had given man his cue. Today most commercial oranges are treated with ethylene gas, which rapidly advances a uniform, appealing gold color.

Further research and automation developed efficient picking and packing. Oranges are carried on huge conveyors through various baths and scrubbings; through wind-tunnels for drying, and through the coloring process, and then moved through sorting, packing, and labeling machines. A far cry from Oriental gardens, where only mandarins could touch the golden fruit.

Perfume from Oranges

Although the sweet orange is the world's most popular fresh fruit, the bitter orange of Seville goes into millions of jars of superior marmalade, flavorings, and candies each year. Fragrant oil for perfume is also pressed from the skins of bitter oranges. Once upon a time the use of orange oils in perfume and drinks was the exclusive prerogative of maharajahs, and the formulas were closely guarded secrets, part of a family inheritance handed down from father to son. Today southern France is the world's perfume center, and orange oils are used on a large commercial basis. The most delicate and fragrant are produced from the blossoms, of which one average bitter-orange tree may produce sixty-six pounds. The matchless perfume made from these orange blossoms is called "neroli" and is exclusive to that area of France.

Twentieth-Century Oranges

In recent decades the advent of frozen orange juice has changed the whole orange economy. The first small experimental laboratories hesitantly began commercial production of frozen orange juice after World War I. It was an instant success, and resulted in the greatest boom any fruit has ever known. Orange juice is popular with growers because it eliminates many problems in shipping, and makes use of large quantities of excellent fruit that would be difficult to market because of irregularities in size, shape, or color. Since World War II, frozen orange juice has become a universal commodity. Even members of the British royal family are said to have made it a part of their traditional breakfast.

What genie of the Arabian Nights could have dreamed up the magic of a modern frozen-orange-juice plant? The juice of countless oranges flows from giant evaporators like water from mountain springs. Condensers, canners, and freezers conserve the very essence of the golden tropics, once produced in royal gardens for the exclusive use of kings, and transform it into the favorite breakfast drink of millions. Mechanical wizardry and modern magic!

PANSY

The pansy, so familiar and so beloved, was unknown to colonial gardens. We tend to think of it as an old-fashioned flower, but it was a late immigrant and did not arrive here until shortly before the Civil War. Even more surprising is the fact that there were no pansies anywhere in the world until the late eighteenth century.

The pansy is a direct descendant of the *Viola tricolora*, the Johnny-jump-up. Most authorities agree that this appealing plant is native to the Austrian Alps, and that it worked its way, like so many other plants, across Europe and then across the English Channel, perhaps as a camp follower of the Roman legions. At any rate the Britons and Celts used it for centuries as a drug, a food, a cosmetic, and a love philter, and by Shakespeare's time it grew wild about the English countryside.

Along the way it had collected some forty local names, most of which have a sentimental reference and must have been prompted by the saucy small face of the Johnny-jump-up that peeped out from herb gardens and woodland paths. Some, like "love-in-idleness," "kiss-her-in-the-pantry," and "peeping Tom," are almost forgotten. But "Johnny-jump-up" and "heartsease" have come down through the years.

As a love philter the *Viola tricolora* was considered such a boon to maidens that "heartsease" became accepted as its official name. As heartsease it was brought to America by the early English settlers, and soon escaped to roadsides and fields to become naturalized here as in England.

Birth of the Pansy

The Johnny-jump-up had found no place in Tudor gardens, but grew only as a useful and interesting wildflower. Then, in the great surge of interest in gardens that occurred during the eighteenth century, its possibilities as a bedding plant were explored.

Work was begun on the Johnny-jump-up toward the end of that century by the gardener of Lord Gambier, a British naval commander who took part in the capture of Charleston during the Revolution. Gambier had been born in the Bahamas, where his father was lieutenant-governor; his garden was one of the most beautiful in England. Perhaps his love of flowers stemmed from a youth spent amidst the lush vegetation of those islands, and gave him the prodigious patience he displayed during his gardener's Johnny-jump-up project. Its aim was to change this small popular peasant of the plant kingdom into a squire that would be welcomed into stately gardens.

Experiments in hybridizing were carried on for thirty

years, during which little Johnny-jump-up refused to surrender his identity. When, at last, a new flower was born—
one that would retain its size and beauty—it bore only a single
strong resemblance to Johnny. The *Viola tricolora hortensis*
had lost the spur characteristic of the violas. Its flowers were
large and flat, with overlapping, often ruffled, petals of contrasting colors that ran a gamut from cream to a velvety black.
But they retained the typical markings of Johnny-jump-up's
amusing face.

Gardeners hailed the pansy, which had all the appeal of
Johnny-jump-up but produced conspicuous, colorful flowers
and leafy green growth. Pansy growing spread quickly
throughout England, France, and the Netherlands, but it did
not arrive in America for another fifty years. The popular
name given the new flower was actually an old one, derived
from *pensée*, meaning "thought," which the French had given
to the small *V. tricolora* centuries ago because of its supposed
power to turn the thoughts of a lover to one's self. When
Ophelia, in her madness, says pansies are for thoughts, she and
Shakespeare are alluding to the heartsease or Johnny-jump-up.

Modern Pansies

The pansy, which still refuses to conform to established patterns and runs in strains rather than varieties, belongs to the
tribe of leafy-stemmed violas. It has become the most popular
of all hardy dwarf herbaceous flowers, and produces best
when grown from autumn-sown seeds. Although it is a flower
of the North Temperate Zone, it will not tolerate hot weather
or bitter cold. But, given cool loam, partial shade, and the
stimulus of bone meal, it will cover your border with a carpet
of bloom.

The modern pansy is almost scentless, but a bed or a
bowl of them past their prime does not make the heart grow
fonder. The scent of decay, which is really a particularly
noxious gas given off by the rotting stems, is the pansy's only

fault. Perhaps it can be thought of as a manifestation of supernatural spirits still inhabiting the pansy. The next time its amusing face looks out at you from the flower border, remember that there is mystery behind that smile.

PEACH

Springtime on the East Coast, the West Coast, and the Gulf Coast, and in the Great Lakes region, is welcomed by mile after mile of fragrant peach orchards. There are approximately 30 million peach trees in the United States, and they produce, on an average, 70 million bushels of peaches annually. Yet not a single peach tree grew in all America when the white man came.

History

The peach was grown in China 2,000 years before the birth of Christ. Ancient Chinese writers called it the "tree of life," and considered its fruit the symbol of well-being and longevity.

It began its long westward journey in caravans that carried it to India and on to Persia, where it became naturalized and where invading Roman legions found it. They called it "malus Persicus," the Persian apple. The Emperor Claudius is credited with establishing it in Greece, which had long since become a Roman colony and an outpost for Rome's eastern Mediterranean expeditions. It soon adapted to Grecian soil, and, as the centuries passed, filled Greek legend and myth with stories of the Persian apple—the golden fruit of the gods.

It flourished in Rome, too, and many a Roman villa had a peach orchard. Virgil, who is said to have been the first European to describe the *malus persicus*, boasts of his peaches in the *Georgics*.

Roman legions carried peach pits to Gaul and Britain. During the long wars of the Middle Ages, peaches and other fruits survived only because monks cultivated them in monastery gardens. In England the endless Wars of the Roses reduced even these outposts of agriculture, and the farms and orchards of many monasteries perished along with the manor gardens.

It was Catherine of Aragon, after she married Henry VIII, who turned the tide. She had a green thumb and a knowledge of horticulture, and she brought peach, plum, and cherry trees into England. She also encouraged the English nobles to compete with one another in the design and maintenance of gardens for palaces and manor houses.

Peaches in America

Peaches were among the first trees the Spaniards planted in Mexico, Brazil, the West Indies, and the California missions. In 1565, Spanish settlers established the first peach orchards in Florida, in Saint Augustine. They flourished over the next two centuries, and became famous; but the British soldiers who captured the town in 1763 destroyed them.

The French brought peach stones to New Orleans and the Saint Lawrence valley, and the ships' lists of English and Dutch settlers show that every colony had its peach trees.

In tidewater Virginia, planters laid out luxurious formal gardens that opened into orchards of peach, pear, plum, and fig. Peaches were planted at Jamestown and Williamsburg, at Boston and Salem, and by William Penn and Lord Baltimore. All colonial areas imported peaches at the earliest possible moment, partly to supply the settlers with fresh fruit and their swine with feed, but more especially as a source of brandy—at least if we can judge by a notice in an early copy of the *Virginia Gazette:* "Planter Mathew Marble advertises sale of land that produces 30 hogsheads of tobacco, 1,500 bushels of corn, 3,000 bushels of wheat, 1,000 gallons of Peach Brandy."

The Indians and Peach Culture

Many Indians were excellent orchardists. They loved the luscious fruit and coveted the peach stones, which they carried westward, planting them about their campsites and along trails from Florida to the Great Lakes. Many a pioneer, crossing the plains, collected the bounty of Indian peach trees.

In 1776 John Bartram explored the hinterland along the Little Tennessee River, where he found abandoned Cherokee villages and noted: "There were also peach and plumb orchards, some of the trees appeared to be yet thriving and fruitful." Years later his son, William, explored the south for the American Ethnological Society. In 1789 he reported that the Creek and Cherokee Indians were excellent agriculturists, with "orchards of peaches, oranges, figs and some apples about their encampments."

But when gold was discovered on the Cherokees' land, the white man destroyed their great orchards, and forced them onto reservations. After all, it was pointed out in extenuation, the British had found it expedient to destroy the peach and apple orchards of the Five Nations as the simplest means of forcing them to surrender control of the Great Lakes region.

Commercial Peaches

One of the first commercial peach orchards in the United States belonged to the Shakers, at New Lebanon, New York. For nearly two centuries the Shakers were the chief source of supply for America's rapidly growing pharmaceutical trade. From their gardens and orchards they created and sold herb extracts, powders, oils, dyes, ointments, confections, and cosmetics, as well as 150,000 packets of garden seeds a year. Among their most popular cosmetics was peach water, a product of their famed peach orchards.

For 200 years after the peach was introduced to America, according to our Department of Agriculture, its culture remained rudimentary. Peach seedlings produced well in the virgin soil, but seldom came true to form. Only a few named varieties were cultivated before 1800, and their propagation was only of local interest.

From 1800 to 1850 the rapid increase in population and the growth of cities on our eastern seaboard inspired farmers to establish commercial peach orchards near Richmond, Baltimore, and Philadelphia, and all of them prospered.

Between 1850 and 1890 our canal and rail transportation systems developed, the west was opened, and pioneering reached its peak. Peach stones and budded seedlings were carried westward by the settlers. New urban centers sprang up, creating increasing demands for commercially grown peaches, and soon refrigerated cars and trucks carried the fruit to market. This led to a demand for peaches that had excellent handling and shipping qualities. Such peaches did not then exist.

Elberta Arrives

Nature is an alchemist who provides new wonders for new techniques of living. At this pregnant moment the Elberta was born—a highly productive and perfect shipping peach, a gift from nature to our growing country.

In 1850, in Shanghai, Charles Downing became so enthusiastic about a white-fleshed, almond-flavored peach he found there that he sent to Henry Lyon, an orchardist friend of his in Columbia, South Carolina, a young Shanghai peach tree in a flowerpot. It caused a considerable stir, because it was the first peach of record to be sent from China to an American orchard. As it grew and bore, Lyon consulted the American Pomological Society, whose specialists recognized that this was a new species of peach, quite different from the yellow freestones they had been cultivating. They named it the "Chinese cling." It became the progenitor of an eventual hundred varieties of modern American peach.

In the fall of 1870, Samuel H. Rumph, of Marshallsville, Georgia, who owned large orchards of Crawford and

Oldmixon cling peaches, was given some pits of the still-rare Chinese cling, and planted them in his orchard. The seedlings throve, and in due time one of the young trees bore, in mid-season, a large peach whose orange-yellow skin was over-spread with a red blush. Its flesh was yellow; its stone was free; it was sweet and tender; and the tree bore more abun-dantly than any young trees of the known varieties. It ap-peared to be a sport with much potential.

Rumph decided that the parent tree of his Chinese cling must have been fertilized by a Crawford peach near which it grew, and that it had produced an entirely new strain of peaches, superior to those borne by either of its parents. He named his new peach "Elberta" for his wife, Clara Elberta.

The ancestry of the Elberta has remained a mystery, and some horticulturists insist that its character is that of a Persian and not a Chinese peach. Whatever the truth is, Rumph's peach has been a winner. The Elberta is now the most widely grown peach in the United States, and the most popular in spite of a few inherent faults. It is, like all peaches, susceptible to frost. Its stone is larger than that of most varieties, and fruit that is picked green may retain a pronounced bitter taste. But these faults, the Department of Agriculture admits, are the price the urban consumer is forced to pay for fresh peaches: white peaches produce less abundantly, and they will not ship, so for a century the Elberta has been the most profitable peach in America, as well as the most popular. It is grown in all the leading peach states, and qualifies admirably for ship-ping, marketing, canning, and eating. Such attributes promise long commercial supremacy.

As a coincidence, another of the original Chinese cling stones planted in the Rumph orchard produced one of the few peaches ever to rival the Elberta's abundant production and popularity. Rumph named it "Belle." The fruit is large, with a red cheek and firm, juicy white flesh, and some experts consider it the finest of all American peaches. It, too, is widely grown.

Resistance to Disease

Added to the Elberta's outstanding qualities are its ability to withstand the attacks of insects and its resistance to the various peach diseases that multiplied as commercial production expanded in America. The peach yellows, one of the first of these to frustrate orchardists, was followed, between 1900 and 1907, by San Jose scale, brown rot, and other fungus infections, and by bacterial diseases that destroyed the trees along with the fruit. Research has shown that in many instances the criminal is the plum tree, both cultivated and wild, which acts as a host and carrier of diseases that attack the peach. In many localities peaches are not grown within a mile of plum trees, and wild plums are eradicated near commercial orchards.

The discovery that peaches and other stone fruit need 50% more nitrogen than apples led to extensive tests at our Beltsville Agricultural Research Center, in Maryland. The test results emphasized that nitrogen is in fact the most important of the nutrients the orchardist must supply. From a half to three-quarters of a pound per tree, applied in the spring, will ensure the tree's health and also guarantee a more colorful skin and therefore more sales appeal, even for the Elberta.

Another essential is a site that protects the trees from frost, wind, and flood. Every tree needs a thousand hours or more of 45° F. weather for growth, and for maximum spring bloom. Yet even Elbertas are killed if the temperature drops too far below the freezing point. So our large commercial orchards are located on the Pacific and Atlantic coasts, and in areas adjacent to the Great Lakes and the Gulf of Mexico. The vast bodies of water temper the air and prevent hard frosts during the peach trees' long dormant season.

Elbertas in the Marketplace

From 1900 to 1950 peach orchards covered one of the greatest areas cultivated in the United States. But the production picture has radically changed, and the ratio of trees to bushels has been completely reversed. In 1900 there were about 135 million trees under cultivation and they produced fewer than 20 million bushels a year. Today there are only 30 million trees of bearing age, and they produce around 70 million bushels.

For this remarkable development we owe the Elberta and the Belle most of the credit. Quite an accomplishment for a pair of sports!

PEANUT

Peanuts were unknown in the northern states until after the Civil War, but baseball made them famous.

When Union veterans became civilians again, baseball was one of the pursuits to which they turned with special enthusiasm. Ball parks opened in towns and villages, and Saturday and Sunday afternoons were devoted to the game. But the spectators needed something to chew on between cheers, and in the minds of thousands of veterans there appeared a vision of the delicious groundnuts they had encountered below the Mason-Dixon line and roasted over their campfires.

Some entrepreneur took the trouble to respond to their nostalgic hankerings. Roasted peanuts appeared at ball parks, and their popularity was instantaneous. Soon the peddlers' carts on city streets carried them as well as popcorn, and they appeared in a variety of forms in candy stores. Thus were peanuts introduced to the damyankees.

History

The first slaves were brought to Virginia in 1619, just as tobacco, rice, and cotton were laying the foundations of an agricultural empire there and in the Carolinas. During the seventeenth and eighteenth centuries the slave trade boomed, and peanuts were an ideal food for the slave ships because they kept well, were nourishing, and were familiar to the Africans. Indeed, tradition says that some slaves, confronting an unknowable future, hid peanuts in their hair, hoping to plant them wherever they came to rest, and so save themselves from starvation. They called them "nguba," the Bantu word for groundnut, which became corrupted to "goober." In the South peanuts are still called "goober peas"—and this is one of the relatively few words of African origin retained in American English.

Peanuts soon became a common food for plantation slaves because they were cheap and prolific—two bushels of seed could produce fifty bushels of peanuts per acre—but this very factor brought them into disrepute as food for the plantation masters. The northern soldiers, however, had no such prejudice. They were often on short rations, and when they were introduced to peanuts and taught to roast them in their campfires—probably by former slaves or camp followers—they discovered a tasty satisfying food they could not forget.

But the peanut, so long identified as an import from Africa, is not an African plant. It is native to tropical America, and was cultivated extensively before the white man came, by the Mayas in Yucatan, by the Incas in Peru, and by various tribes in Brazil, where several species are native. Relics of pottery and sculpture decorated with peanut designs show that these civilizations had used the peanut in a variety of ways since antiquity. The conquistadors carried it to Europe, and the Portuguese introduced it to the coast of West Africa, where they had established a profitable slave trade. It became a common food there long before Virginia was settled.

Botanical Bits

The peanut is not a true nut but a member of the pea family. Its botanical name, *Arachis hypogaea*, is Greek. *Arachis* refers to its hairy stem, which suggests the web of Arachne, the

Lydian girl who challenged Athena to a weaving contest and was turned into a spider. *Hypogaea* is Greek for burying one's head, and refers to a peculiar characteristic of the peanut. It grows from one to two feet tall, bearing yellow flowers not unlike pea blossoms. After the flower is fertilized the petals drop, and the stalk gradually bends downward until it reaches the ground and buries its fertilized pistil, which develops into a fibrous pod containing two or three seeds.

It is obvious that one of the essentials for good production is a loose sandy soil. Otherwise the peanut demands little—an early spring, a hot summer, and soil on the alkaline side. It will thrive anywhere that fills these requirements, and is often grown in flower gardens as a novelty.

Peanut Butter

Although the peanut was not grown specifically for commercial purposes until the 1870's, the postwar demands of northern markets encouraged southerners to ship out what crops they grew. It took a former slave to make peanuts a valuable world-wide commodity.

In 1880 a Saint Louis doctor, seeking a remedy for malnutrition, experimented with peanuts as a source of protein. He ground them up and produced a palatable, digestible item, but the response to it was negligible and it was soon forgotten.

Enter George Washington Carver, born into slavery before the end of the Civil War, who grew up to become a brilliant chemist pioneering in agricultural research. He discovered hundreds of uses for the peanut, not the least of which was peanut butter. After it was introduced to the market, its popularity grew so rapidly that by 1900 one-fourth of the national crop of 4 million bushels was used for peanut butter. Today about three-fourths of the crop is still sold as roasted peanuts and in confections, for the peanut remains a national favorite in those forms, too.

A Universal Crop

Today billions of pounds of peanuts are grown annually around the world. They are an important crop in southern Europe, China, East Africa, and India. They feed millions in Zambia, Nigeria, Ceylon, and Formosa, and in the Moluccas, Pacific islands once dominated by Portugal. Modern Indians in Mexico and Central and South America use peanuts as their ancestors did, for oil, meal, and confections. In our nation, Suffolk County, Virginia, continues as the peanut capital of the south, although for some time to come we may hear less about it than about a Georgia town named Plains.

Peanuts are an invaluable agricultural crop because of their manifold uses. They contain 30%–50% oil, which is used in the manufacture of margarine, cooking oil, and soap, and for industrial purposes. The residue from the oil is pressed into cakes for fattening hogs and cattle. The pods are used as cattle feed; the red skins around the nuts are used as bran. Peanut hay, which averages two tons to the acre, is universally used for forage in field or barn.

In rural areas of Ceylon, peanut leaves are even used for tea. And it seems worth noting that a pound of peanuts contains more protein than a pound of sirloin steak, plus large amounts of carbohydrates and fat.

Finally, and most important, the roots of the peanut enrich the soil in which they grow. Like all legumes, they play a key role in the assimilation of nitrogen from the air, which collects in nodules of nitrogen-fixing bacteria on their roots, and is imparted to the soil for the benefit of other crops. It was George Washington Carver who urged impoverished southern planters to renew their fields, worn out by tobacco and cotton, by planting the benficent and profitable peanut.

Today our south is one of the world's great agricultural areas and the peanut a universal crop. Man recognizes its ability to fertilize and renew the earth while it feeds him and his domestic animals.

PEONY

The peony you plant may bloom for 100 years, and the peony's roots go deep into the past. Man has employed its services and enjoyed its beauty for more than 2,000 years. It is a native of China, where the use of its seeds for seasoning and its properties as a medicinal and kitchen herb were detailed in fifteenth-century records. Then, as civilization developed, the peony found favor with the mandarins, and moved from the kitchen dooryard into the pleasure gardens of the leisured and wealthy. It was called "sho yu," "most beautiful," and also "flower of prosperity," and it wound its way through centuries of Chinese legend, poetry, and art.

The peony had also arrived in the Near East before recorded history. Theophrastus, a philosopher who was a friend of Aristotle and who succeeded him as head of the Peripatetic school, wrote, about 320 B.C., a survey of Mediterranean flora and exotics sent him by Alexander the Great from his eastern expeditions. In this *Inquiry into Plants*, Theophrastus describes the peony under the name "glykoside," but it had acquired its present name by A.D. 77. Pliny the Elder declares that it is the oldest of plants, and an important medicine that cures twenty ills. He also tells us that its name comes from that of Paeon, physician to the gods, who came, in time, to be identified with Apollo. The word was originally used for a hymn of praise to Apollo for his healing power. Dioscorides described the peony as a healing herb.

The Peony in England

The first peonies brought to England by the Roman legions were the so-called male ones, *Paeonia masculata*, which served a medical purpose throughout the centuries of Rome's occupation. They were cultivated as a healing herb in the physic gardens of England's early monasteries, and during the Middle Ages became such an integral part of the island's flora that many believed they were native. Research has proved, however, that peonies growing wild undoubtedly had escaped from monastery or castle gardens.

Some sources say that Eleanor of Aquitaine brought peonies to England from southern France, when she came to be crowned queen to Henry Plantagenet in 1150. She may well have done so; for she did bring rare flowers from her garden, along with such luxuries as perfumes, spices, silks, and velvets. Tradition also says that men returning from the Second Crusade at about the same time carried among their treasures the roots and seeds of the peony for the enrichment of their castle gardens. Legend and superstition obscure medieval history, but two facts are beyond argument: the peony was well established in England by 1200, and it was one of the earliest of the plants brought to England that survive into our own day.

Myth and magic grew up about it during the Middle Ages. The church called it a blessed herb for its healing properties. It was also used as an amulet to protect the wearer from witchcraft and the Devil, and to combat the Evil Eye. It was listed for centuries among other enchanted herbs—monkshood, moonwort, foxglove, and poppy—thought to possess supernatural qualities and to be the most potent for the casting of spells. And such beliefs were slow to die. Even in the eighteenth and nineteenth centuries, in isolated rural areas in both England and America, mothers still hung strings of peony seeds about the necks of infants to prevent convulsions and teething problems and to ease the terrors of epilepsy.

Gerard, who goes to great lengths to dispel these superstitions, provides the first picture of the double red peony published in England. William Turner tells us that the crimson female peony, which was native to Crete and the eastern Mediterranean, was brought to England from Flanders by wealthy Huguenot refugees. He adds that this female peony (a term used to distinguish the Western varieties of *Paeonia officinalis* from the *P. masculata*) is confined to the gardens of the rich, because it costs twelve crowns a root.

Sir Francis Bacon, in his famous essay "Of Gardens," lists the double peony among the April flowers, along with "flower-de-luces, and lillies of all natures," tulips, daffodils, and lilacs—exotics from foreign lands. In 1629, at Kew Gardens, Parkinson was growing the peony Byzantium, which he says was introduced from Constantinople.

Peonies were foremost among the plants that began pouring into Europe after the establishment of the Dutch and British East India companies. More than 100 varieties were then grown in the gardens of the East, and many became popular in England. John Hill, an herbalist-author of the period, wrote that there was not a flower more known than the peony in English gardens. Peter Pallas, a German naturalist who became a great favorite of Catherine the Great, and the first foreigner to explore the plants of Russia's hinterland, corresponded with many English botanists, and sent seeds and plants to Joseph Banks, who grew them at Kew Gardens in 1784. Among these was the lactiflora peony, which was then considered an edible species. Pallas reported that the Mongols used the roots for a common soup, and people throughout Siberia flavored their tea with its ground seeds.

In the following decades many beautiful varieties were developed in England. Old strains of lactiflora are the pink Fragrans, the white Whitley, and a dark pink Humei, still found in long-established gardens. These and the early red *Paeonia officinalis* (*rubra plena*) form the basis of early hybrid crosses.

The Tree Peony

The Dutch East India Company brought the first tales of the tree peony to Europe, tales that so fascinated young physician-botanist Engelbert Kaempfer that he enlisted as surgeon with the Dutch fleet. As soon as he saw the tree peony he knew that its fabled beauty was unique and that it must be priceless. Indeed, its value could scarcely be exaggerated, for one tree of a choice variety might bring 100 ounces of purest gold. Considered a rich inheritance because they lived for 200 years, they were closely guarded in the mandarins' gardens, and were included as an important item in marriage settlements.

A vast amount of Chinese literature on the tree peony accumulated through the centuries. It had come into its own during the T'ang Dynasty (A.D. 618–906), a period famous for conquests and extraordinary cultural burgeoning. Records of this time show that thirty-nine varieties of the peony were grown, many in the gardens of Buddhist monasteries. The Buddhist monks had carried the tree peony to Japan in the fifth century, and it was from Japan that Kaempfer shipped the first specimens to Europe.

The tree peony is a perennial woody shrub that belongs to the Ranunculaceae, or buttercup, family, as all the peonies do. It grows from six to eight feet high, and may be covered with 100 blossoms at one time, each six to seven inches across. A heady fragrance and varied color add to its appeal. It is lovely in winter, too, retaining its shape, and it puts forth new spring leaves from the persistent stem, rather than from the ground, as the herbaceous peony must do.

Kaempfer's explorations inspired him to write *Amoenitates Exoticae.* (See Hydrangea.) His watercolors in it of the tree peony were the first to be seen by the West, and its beauty created a demand for it from European gardeners.

But it proved a sensitive shrub not adapted to transportation or quick reproduction, and seldom came into its full beauty in Western gardens. It also remained rare and expensive: even in 1830, one tree peony cost six guineas.

Joseph Banks brought the first tree peonies to Kew Gardens in 1771, and by 1800 they were also known in France.

Robert Fortune (see Forsythia) sent tree peonies to England in 1846 and 1851, and in 1912 William Purdon, a late collector, sent to Kew a new tree peony with dark-red flowers that he had found growing wild on the borders of Tibet. Its seeds were sent to the Arnold Arboretum, where E. H. Wilson introduced it as *Paeonia veitchii* in 1913.

Purdon also explored Peony Mountain in Kansu province, where Chinese tradition said tree peonies grew wild in such abundance that they perfumed the air, and for centuries had been used as firewood by the local villagers. Perhaps this is the reason Purdon found no tree peonies there in 1912. Reginald Farrar, another English collector, discovered a magnificent wild white one in southern Kansu two years later. But the wars and revolutions that have troubled the world almost continuously since 1914 have reduced the possibilities for plant discovery, and botanists have turned to other aspects of their calling.

Victor Lemoine, in Nancy, took up the hybridizing and development of the tree peony. He increased the blooming season, the range of color, and even the size of the flowers— work furthered by the late A. P. Saunders, of Hamilton College, who strengthened the flower stems so that they would hold the great heavy heads upright and do justice to their bounty and beauty.

Hybrid tree peonies are generally derived from the larger Chinese peony, *moutan*, and the *P. lutea* from Japan, which is longer-flowering and has a wider range of color.

Peonies in America

The tree peony had been introduced to America directly from Japan in 1891, but specimens of the old Chinese tree peony in Robert Fortune's collection had been imported from England and grown on a few estates before the Civil War. Its exorbitant cost and its sensitivity to the cold generally limited its spread in the United States. There have been periods, however, during which there was something of a craze for tree peonies here. Possibly a new wave of interest is upon us, inspired by various spectacular hybrids now available from some American nurseries. With reasonable care these hybrids will survive winters in southern New England and westward in the same weather zone.

As for the herbaceous peony, it has been in the United States since earliest colonial days. It was brought to Virginia in the 1600's, and was a prized introduction in Williamsburg gardens in 1698. Jefferson's garden book records that the peony was planted at Monticello; Washington grew peonies at Mount Vernon; and John Custis took pride in cultivating several varieties.

John Bartram is said to have exhibited peonies in his botanical garden, and they were in such high fashion in other Quaker gardens before the Revolution that at the Philadelphia Centennial in 1876 the peony was featured as the flower that symbolized the American spirit and ambition and determination to adapt and thrive.

The glory of New England's front yards was the red "piney," as it was commonly called in rural America. No old-time garden was thought complete without its neat foliage and incomparable blooms, and it went west with the pioneers in thousands of covered wagons. Like the pioneers, it conquered the continent, and it thrives everywhere but in the Deep South.

Salem and Portsmouth claimed to have peonies a century old when the first clipper ships brought in a flood of new peonies from China and Japan. These included new varieties of the lactiflora, and the tenuifolia, with its feathery foliage.

Today the American Peony Society, a nationwide organization of peony growers, offers information and service to novice and specialist alike. There are literally hundreds of varieties available, and they come in virtually every color and shade except pure yellow. Even the mythical blue peony, which the Royal Horticultural Society ordered Fortune to find on his first China assignment, now has many rivals in modern hybrid shades. There are singles and doubles, fragrant japas, the anemone-flowered peony, the semirose, the crown, and the bomb—an endless choice.

Few plants offer your garden so many advantages as the herbaceous peony. It is long-lived and extremely hardy; it will endure partial shade and indifferent soil; and it is disliked by slugs, bugs, mildew, rabbits, and most plant diseases. It does love rich well-drained humus, lime, or the old-fashioned wood-and-coal-ash treatment, and appreciates shallow planting. It deserves a permanent place in your perennial bed, where it can reign in regal beauty.

PETUNIA

Twenty-odd years after Columbus explored the Caribbean islands, Juan Díaz de Solís, a Spaniard seeking a southwest passage to China, sailed into a splendid estuary on the east coast of South America, thus discovering the Río de la Plata. His soldiers, trampling underfoot the scrubby white-flowered weed that covered the shores of what is now Uruguay, were probably the first Europeans to see a petunia. But there is no record that they paid any attention to it—and small wonder. Solís and several of his companions were killed by the Indians.

Ten years later, in 1526, Sebastian Cabot, sailing for Spain on the same mission, anchored in the same harbor. Cabot understood its tremendous possibilities, especially because the Indians, who appear to have been friendlier than those who dispatched Solís, reported that the distant hills held fabulous stores of silver. He built a fort above the river, to which he gave the name it has born ever since, and left some soldiers to hold this Argentine—this land of silver—for Spain.

Greed for gold and silver so preoccupied the Spaniards that it did not occur to any of them that the scrubby weed covering the fields and riverbanks would one day be the Argentine's most widely distributed export.

Three hundred years passed—three centuries of Spanish rule, which meant the cruel subjugation of the Indians and the mining of their silver. It was not until Spain fell to France in the Napoleonic Wars, and Napoleon placed his brother Joseph upon its throne in 1808, that Argentina won its independence. A French commission that was sent to Argentina in 1823 to evaluate its resources included several amateur naturalists. They became interested in the weed that grew so abundantly along the Río de la Plata, which the Indians scathingly called "petun," a worthless tobacco. But the French thought it might hold the possibility of an economic bonanza; they sent its seeds to Paris, to the distinguished botanist Antoine de Jussieu, and asked him to classify the plant and offer an opinion on its value.

Jussieu confirmed the Indians' identification, but events confirmed their low opinion of what was indeed a dwarf tobacco. It belonged, therefore, to the Solonaceae, or nightshade, family of plants, to which the tomato and potato also belong. (See Potato.) These other relatives had contributed so much to European economy that the possibilities of the *petun* were eagerly considered. But no significant use emerged for the creeping minor tobacco, which was given the Latinized name *Petunia nyctaginiflora,* and finally recommended as a new plant for the flower garden. Few European gardeners, however, paid any attention to it.

Argentinian independence had an immediate impact on Europe. The nation established the world's most liberal immigration laws, opened her doors to experienced farmers and skilled labor, gave large tracts of land outright to settlers, and subsidized farm equipment. Among the talented men who emigrated to the Argentine was James Tweedie, head gardener at Edinburgh's botanical gardens, who seized the opportunity to explore the natural resources of a vast region that had been closed to foreigners for 300 years by the edicts of the Council of the Indies.

Tweedie arrived in Buenos Aires in 1825, and for the next thirty-five years collected new and rare plants, which he sent to botanical gardens in Scotland, England, and Ireland. Sometimes on horseback, sometimes by boat, but often on foot, he explored the area from the snowy Cordilleras to the Atlantic coast, and from Patagonia to the Río de la Plata. He collected many treasures, but his most fortunate find for gardens was that of the *Petunia violaceae*, in 1831. It was a low-growing trailing plant with a small, dull, mid-purple flower, markedly different from the scented white petunia, *P. nyctaginiflora*. Both varieties first appeared in English gardens about 1860. Although their flowers were still small and inconspicuous, they were produced with an abundance that made them widely appreciated in Europe. But they were not grown in America for half a century more.

American Hybrids

California nurserymen were among the pioneers in hybridizing petunias. Their success was soon rivaled by that of seed houses in other states, and American gardeners responded to the endless procession of hybrids with an enthusiasm that was presently echoed in Europe. Few garden flowers have enjoyed a greater ultimate triumph. By now there are single and double petunias, ruffled and fluted ones; there are maxis, multifloras, and balcony hybrids that tumble over walls and out of flower boxes and pots; there are fragrant petunias and night-scented hybrids that suggest their relative the garden Nicotiana. And the color range, from white to deepest purple, now includes not just brilliant red, but a clear yellow as well as salmon and orange shades. Each type has its own special excellence, but although these hybrid petunias are strictly modern, they all derive from the original white *P. nyctagini-flora* and the purple *P. violaceae*, crossed with another strain, the *P. bicolor*. None will come true from their own seed, and they all tend to revert to the old purple and white, so most hybrids continue to be propagated by greenhouses.

Some seed houses sell tons of petunia seed. When one learns that there are 300,000 of them to an ounce, and that actual tests have proved some hybrids bear 500 blossoms in a season, the petunia's prodigality staggers the imagination. It spreads its gay flowers over the modern world. It will grow in every state of the union, including Alaska and Hawaii; it will grow in pots or plots, beds or borders, small gardens and city parks. And I myself have followed it across Europe: from Italy to Holland it bloomed brightly in flower beds, and in window boxes on the balconies of farmhouses, chalets, and city homes.

New Techniques in Hybridizing

One of the newest techniques of hybridizing involves the introduction of high-intensity light in the growth chambers of nurseries. This speeds up the reproduction, and the results are fantastic—in actual experiments petunias have grown from seed to flower in thirty-seven days. This is a tremendous boon to commercial growers, because the process takes nature from, at best, 90 to 100 days in the average garden.

The Year of the Petunia

Nineteen-forty-four was the "Year of the Petunia." That summer a severe drought killed great quantities of flowers in gardens across the country, but the petunia, thanks to its weedy endurance and love of the sun, came into its own. It bloomed so abundantly, in so many sections of the country, that it became national news, and established its popularity for all time. Gardeners even discovered that petunias made long-lasting bowls of cut flowers, and that they would continue to bloom gaily if potted and moved indoors in the fall.

A Petunia Rental

Once I summered near Philadelphia in an old Quaker house with a beautiful garden. The only rent took the form of a surprising but understandable request. I must pick the petunia flowers every day so the plants would not go to seed but would still be blooming abundantly when the owners returned in the fall. They added, smiling, "You see, we love petunias."

PINEAPPLE

Pineapples have become virtually synonymous with Hawaii, but they were well known for three centuries before they reached those islands.

The pineapple's written history began one day in 1493, when Columbus sailed into a harbor of the island that would be known as Guadeloupe. He called the lovely spot Puerto Bello, and wrote in his journal that a well-laid-out village spread around the harbor, with gardens of vegetables surrounded by fields of maize, and by orchards of a fruit no white man had seen before. In his report to Ferdinand and Isabella, he described it as "resembling green pine cones, very sweet and delicious," and said that trees and clefts of rock were found "dripping with an abundance of its honey."

He called the fruit "piña de las Indias" but the Indian name was "ananas," "fruit of excellence." Strangely enough, most Europeans still use some version of this Indian name. Only English-speaking people call it "pineapple," although *piña* is used in Spain, *piña blanca* in Latin America.

The Indians believed that the ananas had been brought to Guadeloupe from the Amazon many generations before by the warlike and ferocious Caribs, who long dominated the Lesser Antilles, and whose custom it was to bring home seeds, roots, and plants from the places they invaded with their great war canoes. Later research proved that the people of the Amazon had indeed cultivated the ananas for centuries. It was as old a food as the potato—pineapple-shaped jars have been found in pre-Inca graves—and it had become an essential plant in the West Indies by the time Columbus arrived there. It was a delicious food, it produced a potent liquor for festive occasions, and it was used as a poultice to reduce the inflammation of bruises and other wounds. Most important to the Carib warriors, the decayed fruit yielded an effective poison for their blowguns.

All this was reported by Peter Martyr, court historian to their Spanish majesties, who collated the official reports of the first Spanish explorations, and published a collection of 816 letters from early conquistadors, describing their discoveries and discussing contemporary events.

The Spanish found the pineapple an ideal fruit to send home for cultivation, and few other New World fruits have had such an enthusiastic initial reception in Europe. Within a few years Portuguese, Dutch, and Spanish vessels had carried the *piña de las Indias* around the world, and by the end of the sixteenth century it was growing in India, China, and the East Indies. It was peculiarly adapted to migration because its stiff crown of leaves retained moisture, enabling it to survive the long voyages and to grow in semiarid regions. In today's world markets it is second only to the banana as a commercial tropical fruit.

The Europeans' enthusiasm for the pineapple seems not to have been dampened because their climate was unsuited to it. Although few of them had an opportunity to taste or see it, their interest was kept at a high pitch by continued reports from missionaries, travelers, and explorers who had eaten it, and by early drawings of it. A Hollander named Georg Eberhard Rumpf was the first to depict "the heathen fruit," which he called "Bromelius comosa" and described in both Latin and Dutch. John Parkinson, who included the pineapple in his encyclopedic English herbal *Theatrum Botanicum*, or

The Theatre of Plants, published in 1640, entitled his illustration "The West Indies Delitious Pine."

Oliver Cromwell is credited with having introduced the first pineapple into England. In 1655 the captains of a fleet that he had dispatched to the Spanish West Indies captured Jamaica, and on their return to England they presented Cromwell with a *piña de las Indias* as a token of their victory. (See Fuchsia.) Not to be outdone by an upstart Puritan, Charles II, soon after his restoration to the throne, saw to it that he was presented with the "first" pineapple, brought to England by Matthew Decker, an ardent royalist. The king's gardener, aptly named Rose, offered it to him on bended knee, and Charles had a painting made of the whole affair to establish his position as "Patron of the Pineapple."

The intense public curiosity about this fruit from the West Indies continued into the next century, and in 1720 a life-sized portrait of a pineapple was hung in the Cambridge Museum for the public to view. It may be seen there today.

Glasshouses

About 1700 a merchant in Leyden named La Cour had devised a glasshouse in which he grew to maturity a pineapple that was sweet, fragrant, and of good size. The event caused great excitement: it opened exhilarating new possibilities for gardening, and a vogue for glasshouses sprang up in England and on the Continent, not only for the commercial production of flowers and food crops but also as an adjunct to every estate and country home of any consequence. Numerous publications describe the construction of glasshouses and techniques for growing plants in them. They became a hobby and a status symbol, and glasshouse pineapples became a popular dessert for the privileged.

By 1861, when Mrs. Beeton's *The Book of Household Management* first appeared, glasshouse pineapples would seem to have become generally available—at a price. After providing a glimpse of pineapple history and a tempting recipe or two,

Mrs. Beeton advises her readers to make preserves when "foreign pines" are in season—July and August—inasmuch as they are cheaper than English glasshouse pines: she lists the going price as one to three shillings each for foreign, but from two to twelve shillings per pound for the glasshouse fruit.

Modern shipping and refrigeration have generally outmoded glasshouse pineapple culture. But it continues to thrive in the Azores, where it constitutes a major industry: glasshouse pineapples from the Azores are shipped ripe, sweet, and fragrant to many European markets, and sold as a luxury.

Pineapples, a Symbol of Hospitality

Columbus had found that the Indians hung crowns of pineapple leaves on village gates and at the entrance to their homes as a sign of abundance and hospitality within. The Spanish copied this pleasant custom in their New World homes, and carried it back to Europe, where it became widely popular. Pineapples were carved on doorways and gateposts, and cabinetmakers seized on their possibilities, decorating bedposts, desk finials, chests, and chairs with them. The vogue spread throughout Europe, and a century later the pineapple design had found its way back to America and was used throughout the colonies by architects and cabinetmakers.

In tidewater Virginia the pineapple symbol of southern hospitality can be found today on the river landings and house gables of old plantations. Among these is Brandon, the estate owned by the descendants of Captain John Smith's friend of that name, to whom it was granted in 1616. Brandon's pineapple piers are a nostalgic reminder of an era when rivers were the main roads in America, and guests arrived in long canoes manned by singing paddlers.

Hawaii at Last

Although there are records of the pineapple's having been introduced into many islands and many nations, and of the pineapple "seeds" that were gathered in Brazil by Captain Cook and Joseph Banks and planted in Tahiti in 1769, there are no records of when or how they came to Hawaii. The most generally accepted theory is that they either floated in from a wrecked Spanish or Portuguese ship, or were brought ashore by sailors, and discarded. This is reasonable enough, because we know that both the Spanish and Portuguese carried pineapples on their sailing vessels to prevent scurvy, just as the English carried limes.

Whenever it arrived, the pineapple found an ideal environment in Hawaii, and flourished there. The first white settlers discovered it growing wild beside the sugar cane the Polynesians had brought from their native islands centuries before, and planted as hedges about their fields and huts.

Such abundance aroused the commercial interests of the white men. Sugar plantations, and, later, pineapple plantations were set out. But there was a scarcity of labor, because the Polynesians, who had led an idyllically easy life in their island paradise, did not wish to labor all day in the fields. And the water supply was limited. But engineering soon took care of irrigation, and the labor problem was solved by importing contract workers from China, Japan, Puerto Rico, and the Azores; from Mexico and southern Europe; and from elsewhere in the South Pacific.

Hawaiians watched while thousands of laborers of every race and nation poured into their islands and the ever-growing sugar and pineapple plantations claimed their land. Together these two industries, created and dominated by the white man, have changed the face of the islands and the faces of their people. (See Sugar.)

Pineapple Cultivation

What appears to be the first written reference to pineapple culture in Hawaii is an entry in a diary kept by the Spanish counselor to King Kamehameha, Francisco de Paula y Marin, who mentions the planting of pineapples and oranges on January 21, 1813. The first attempts at commercial planting were on Oahu, where fifty acres of wild Kailua pineapples were set out about 1840. These prospered well enough to inspire attempts to export the fruit. From 1849 to 1850, 12,000 pineapples were shipped to California, where thousands of hungry men were pouring into the gold fields. Despite the welcome they received, the venture was not a financial success because many spoiled on the way; but the growers decided that all the pineapple needed to gain a wide export market was advertising and better shipping methods.

America's first world's fair, the Philadelphia Centennial Exposition of 1876, provided a perfect opportunity to acquaint a wide public with the Hawaiian pineapple. Visitors to the exposition encountered not just the fresh fruit but canned pineapple as well—a true innovation, since commercial canning was in its infancy—and appear to have relished it in both forms. The pineapple entrepreneurs determined to develop more and better pineapples and to perfect the canning process.

Ten years passed, and then a curious genius arrived in Hawaii. He was Captain John Kidwell, a gifted English botanist, who brought with him 1,000 "smooth Cayenne" pineapples from the West Indies, to improve the quality of Hawaii's wild Kailua pines. He considered the islands' climate ideal for his experiments in the development and cultivation of the pineapple. Today Kidwell's cayennes are the basis of a multimillion-dollar business that employs 22,000 people annually. They are the only variety grown commercially.

In 1898, during William McKinley's administration, Hawaii—which for the previous four years had been an independent republic—became a territorial possession of the United States. Before the year was over, a ship from California carrying hundreds of settlers who had been granted Hawaiian homesteads docked in Honolulu. Their arrival coincided with that of a young Yankee promoter named J. D. Dole, who almost instantaneously decided that Hawaii's commercial future lay chiefly with pineapples. He persuaded the homesteaders to plant their land to pineapples, and he himself joined forces with other energetic and ambitious Americans to develop commercial canning and shipping. Thus was a great industry born.

Mechanized Agriculture

Today pineapples are grown on all the islands, but commercial crops are centered on Lanai, often called "Pineapple Island," where reservoirs and great wells 1,270 feet deep provide water for irrigation, and 15,000 acres of pineapples grow on the western plateaus, 2,000 feet above sea level.

After plowing, machines cover the fields with strips of paper mulch to conserve moisture and heat in the soil and retard weeds. The pineapple slips, suckers, or crowns are still planted by hand in holes in the paper, spaced three feet apart. Then a mechanized mother boom takes over. Its great arms, sixty-five feet long, spread over the fields. Besides watering, it sprays: with herbicides to kill the weeds, with insecticides to kill the bugs, with fertilizers to increase growth, with hormones to regulate ripening, with iron sulphate to control acidity in the soil. Under this mechanical maternal care, each pineapple plant grows two to four feet high and bears one fruit, which averages three pounds.

The first harvest begins from twelve to twenty months after planting. Shoots along the stem bear later crops, and commercial growers harvest two to five crops from a planting, but each year the fruit grows smaller. Old plants are torn out, and disintegrated to fertilize the field. The crowns, slips, and ratoons—new shoots from the roots—are replanted, and the cycle is repeated.

The pineapples ripen unevenly, so there must be successive pickings, which, like the planting, are done by hand. But research is seeking ways to mechanize even these prerogatives of man.

Each year Lanai ships 122 million pineapples to the canneries in Honolulu by barge. During the peak of the season, 6 million a day are processed.

Behind the Can

The first experimental canning of pineapples is said to have been accomplished on a kitchen stove, but then Captain Kidwell stepped in and designed the first successful tin can. With the aid of his partner, a plumber named John Emmeluth, who made the cans, and with promotion by J. D. Dole, who urged co-operative advertising on the growers, the industry was on its way by 1900. That first year, 1,800 cases of pineapples were canned. Today, about 31 million cases of fruit and juice are produced annually.

Modern canneries are models of mechanization, too. Conveyor belts move the fruit through washing, sorting, coring, and peeling, all accomplished by one fabulous Ginaca machine. The fruit is sliced, diced, juiced, frozen, canned, syruped, sealed, cooked, cooled, labeled, boxed, and the boxes roll off conveyor belts ready for shipment to your grocery shelf. Eighty percent of Hawaii's pineapple crop is sent to mainland U.S.A.; the rest reaches world-wide markets.

Today Hawaiian pineapples compete with pineapples from many parts of the world, including Africa, Australia, Malaysia, Mexico, and the East and West Indies. Many varieties have been developed to fit varied soils and climates. Air shipments are becoming economical, making fresh pineapple a rival to canned. Picked, packed, and flown right from the field, these pineapples arrive in city markets at the peak of fragrance and taste.

By-Products

The versatility of the pineapple creates an ever-expanding industry. Science and research have developed fertilizers from harvest residue to enrich the fields; silage, bran, and hay for the livestock; pineapple vinegar and alcohol for people. New fibers made from pineapple leaves are woven into beautiful piña cloth, a most successful venture in the Philippines. Cloth from cruder grades is used for draperies, linen, and bags. Scientists have also searched for the ancient medicinal use of pineapples and developed bromelin, an enzyme isolated in the stem, which is used in the manufacture of drugs and meat tenderizers.

The Unique Pineapple

The pineapple, unique in its structure, is the only cultivated fruit whose main stem runs entirely through it, terminating in a crown of leaves that bears the viable bud of a new plant. Shining like a jewel in a coronet, this bud appears when the fruit is nearing maturity, and indicates that it is ready to reproduce itself. Below the crown, in the green shell of the pineapple, row on row of light-blue abortive flowers appear, fulfilling their purpose in about two weeks. When the last floret has fallen off, the shell is covered with the eyes of the individual plants that grow around the main stem; these develop into a compound fruit made up of many individual ovaries. As it ripens, the shell turns a golden brown, the eyes grow smooth, the fruit becomes sweet and juicy. The starch in the stem is now changed to sugar, which increases 100%. Minerals, acids, proteins, and sugars all blend to make the pineapple uniquely nutritious, delicious, and fragrant.

All pineapples belong to the Bromeliaceae family, as does their surprising cousin Spanish moss—which is known in Hawaii as "Dole's beard."

The Pineapple's Meaning for Moderns

The pineapple has created a $150 million industry in Hawaii. It has utilized the ideas of many men of diverse minds. It has created a cosmos of international living with labor from many nations and races. It has replaced the primitive huts of field hands with model housing, and the "man with a hoe" with machines—all this in our century. We may, if we care to, see a lesson here. Mechanized labor and model housing provide physical ease but create a spiritual void. Man withers when he is separated from the earth, for his roots, too, like those of plants, are in the soil. Labor with the soil is not toil, but tribute, a source of growth and pleasure.

PLANTAIN

Every summer thousands of Americans dig millions of plantains out of their lawns. Yet until our modern age this abominable and persistent weed was considered a valuable medicinal herb and an edible green for the stewpot. Plantain was grown in every monastery garden and cultivated in botanic gardens—and it was raised for bird feed, as in fact it still is today.

The familiar broadleaf plantain, *Plantago major*, is a Eurasian weed of ancient origin. Its tough taproot, its flower spikes bearing countless seeds, and its broad round leaves, which crowd out competitive plants and grass, combine to give it the aggressive character with which it has attained world-wide distribution.

Plantain was growing in England before recorded history. It is recommended by Alfric's *Lacuna*, an Anglo-Saxon leech book of the eleventh century, as "waybreade," a beneficent herb that grew along roadsides and offered its smooth cool leaves for the relief of the footsore, a common affliction in an age when walking was the main mode of travel. In the Middle Ages it was often called "follower-of-the-heel-of-Christ" by penitents who made pilgrimages on foot to great religious centers. It appears as a holy herb in many pictures of the Nativity, and Albrecht Dürer, the most influential artist

of the German Renaissance, often used plantains and dandelions in the foreground of his woodcuts, engravings, and
paintings.

In English literature the plantain held a high place, too.
Chaucer refers to it as a familiar rural herb, and Raphael
Holinshed, whose sixteenth-century *Chronicles of England,
Scotland, and Ireland* Shakespeare drew upon for his historical plays, speaks of it as a household cure-all familiar to
everyone. When Romeo urges "a plantain for your broken
shin," he is referring to a property ascribed to the plantain for
centuries. It was also used, particularly in southern Europe,
for scorpion bites and snake bites, from which it acquired the
folk names "snake plant" and "serpent tongue." And it was
long used as the most effective cure for the deep cuts from
scythe and sickle farm workers suffered at harvest time.

The medicinal part of the plantain is in the broad, ribbed
leaves; these contain a soothing, mucilaginous fluid so effective that merely bruising them and applying them as a
poultice to a wound or festering flesh may bring relief.
Gerard tells us that the juice of plantain dropped in the eyes
"cools the heate and inflamation thereof," and that the reason
for its fame is its "great commoditie" of growing everywhere.
During Tudor days it won favor among the English nobility
as an "effective easement" for gout, an affliction from which
Henry VIII suffered grievously.

Plantains in New England

Plantain was yet another of the herbs the Puritans brought to
New England. It grew in Boston and in Plymouth and in the
early settlements on Cape Cod. In Saybrook, Connecticut, the
wife of George Fenwick, patentee of the colony, led a lonely,
homesick life. But she was a gardener of note, and consoled
herself by laying out an herb garden, in which it is said she
cultivated plantains. We may safely assume that she did so—
plantains grow rampantly today among the traces of the great

house, the garden, and the abandoned grave of the unhappy Lady Fenwick.

Plantain seed was soon discovered by American birds; aided by the wind and by settlers' boots, it spread so rapidly in the virgin soil of fields and roadsides that the Indians said it sprang up wherever the settlers trod and they gave it the name "white man's foot." Henry Wadsworth Longfellow, in the *Song of Hiawatha*, causes his hero to lament that the plantain presaged destruction of the wilderness. Peter Kalm, one of the first European botanists to report on American gardens, observed in 1798 that plantain was so common everywhere in New England that he supposed it must be native to the area.

Englishmen took the plantain to all the Atlantic colonies, and it was used as a home remedy for deep cuts, festers, and sore feet for generations before any country doctors set up a practice. Thus encouraged, it pushed westward with the pioneers, conquered the continent, and became a despised and troublesome weed.

But its healing properties are not altogether ignored, even today. It is noted in *Kiehl's Botanical Handi-Book*, which says the leaves and roots of the plantain are used in ointments, are opposed to putrefaction, and are a specific against scrofula. And people living in isolated parts of Appalachia and perhaps in other remote rural areas use it much as the early New Englanders did.

Plantains in Asia and the Antipodes

As the English established themselves in the Far East and the South Pacific, in the seventeenth and eighteenth centuries, they carried the plantain there, also. It was soon widespread in Australia and New Zealand, where the natives unknowingly echoed the American Indians and called it "Englishman's foot."

It was one of the first European herbs grown in the

Calcutta Botanic Garden, which the British East India Company established in 1786 as a plant repository and distribution center. In 1813 dandelion and clover also grew there among the medicinal herbs—perhaps to relieve homesickness, a frequent disease of Englishmen in India. In China and Japan, it was used for centuries—and still is—as a native, edible green.

Selected Weeds

Selected Weeds of the United States, the illustrated handbook of our Agriculture Department, says that broadleaf plantain is a native perennial of Eurasia that has become naturalized throughout the United States. It also lists the lanceleaf plantain, *P. lanceolata,* a Eurasian annual distinguished by its narrow leaves and the short cylinders of seed at the apex of its flower spikes, from which it takes its common name, "buckhorn plantain." Without offering any service to man, it has invaded every state, and is particularly troublesome in the grasslands of the prairies and the high plains.

A third variety, *Plantago rugelii* or blackseed plantain, thought to be a native perennial, is found only in the eastern half of the United States. Its appearance is similar to that of the broadleaf plantain, but it may be distinguished by its large black seeds, its wavy round leaves, and its leaf stalks, which are tinged with purple.

In defense of plantains, both Henry Ellecombe, a nineteenth-century English gardener, and the famous American naturalist John Burroughs comment that its broad leaves preserve the earth's warmth and moisture, and provide a tempered climate for the seedlings of more delicate plants.

So the plantain demands a modest tribute from those who dig it from their lawns, if only in the knowledge that it was brought to America to serve man, and that both wild birds and those in cages still covet its seeds.

POINSETTIA

In 1825 the United States sent Dr. Joel Roberts Poinsett, of Charleston, South Carolina, as our ambassador to the newly independent Republic of Mexico. This was no ordinary assignment, because Mexico was in the midst of the half century of revolution that followed three centuries of Spanish rule. Nor was Poinsett an ordinary ambassador, but a famous plant collector. Mexico's flowers were little known in the United States, even by the early nineteenth century, and Poinsett rejoiced in the opportunity to explore its green world. If his journey to the capital was fraught with danger from the unsettled populace and roving bands of revolutionaries, this only added to the challenge of an exotic country with a bewildering abundance of unknown plants.

As his horses picked their way along the ancient roads, the sun sparkled on mountain streams and unfamiliar vegetation. There were great trees covered with flowers, some like stars and some like golden balls, and others that bent under clusters of lilac and purple; and there was jasmine everywhere, trailing its fragrant blossoms from on high. Most dramatic of all was a tall shrub that grew along the roadway and in the patios of old churches. It bore great crimson flowers, and the Indians called it "Nativity flower," *flor de la noche buena*, because it blossomed at Christmastime and its holiday greens and reds were used to decorate churches and shrines.

Poinsett had never seen anything like it, but it appeared to be a crimson spurge, a beautiful euphorbia. He pictured it growing in his famed gardens at Charleston, and at his plantation in the Carolina countryside, Casa Bianca, where he had collected plants from the ends of the earth.

Poinsett remained our ambassador to Mexico until 1829, and in the course of his four years there he won wide recognition for his knowledge and love of plants. He in turn observed the Mexicans' love of flowers, their extensive use of native plants for medicine and dyes, and their skill in cultivating sugar, coffee, bananas, oranges, and grains—the economic crops that Spain had introduced to Mexico through the centuries. This mutual interest enabled him to establish a plant exchange, and despite the spasmodic civil war that kept the countryside in turmoil, he introduced a number of plants and trees to Mexico, including the American elm; the Indians responded by helping him to add to his own plant collections.

He sent many seeds, plants, and bulbs to the United States, not only to his own gardens, but also to various friends, particularly to Robert Buist, a Philadelphia seedsman. The crimson spurge was one of the first shrubs to arrive; Buist recorded that by 1828 it was thriving in his greenhouse. He recognized the commercial possibilities of this dramatic plant, which bloomed at Christmastime in Christmas colors, and which propagated easily. He soon offered it for sale in Philadelphia and New York, and in a few decades it was being sold widely.

European nurserymen begged Buist for it, and before long it was growing not only in their greenhouses but in the Mediterranean areas as an ornamental.

Naming a Beauty

Such popularity demanded a more specific name than "crimson spurge." It was a happy coincidence that our first ambassador to the Republic of Mexico was a plant collector, and our good fortune that the Indians' Nativity flower was among the first of its native plants to be directly imported to the United States. Coincidence again played a part in its story when William H. Prescott, one of the greatest of American historians, was asked to give it a name. He had just published his *Conquest of Mexico,* and he knew that country well. He was also a gardener of note, a friend of Poinsett's, and deeply appreciative of the valuable work he had done in Mexico as both plant collector and diplomat. So he suggested the crimson spurge be named "poinsettia," in Dr. Joel Roberts Poinsett's honor.

Botanical Bits

The poinsettia belongs to the Euphorbiaceae family, a widely distributed group of herbs, shrubs, and trees. Its botanical name, *Poinsettia pulcherrima,* attests to its singular beauty. It blooms from November through January, and in its wild state attains a height between eight and twelve feet. Six to eight inches in length, its bright green leaves have pointed, shallow lobes. Its bright-colored bracts look like great scarlet flower petals, but the true flowers are negligible yellow ovals at the center of the red bracts; they contain bristling bunches of stamens and pistils.

Commercial florists have reduced it with growth retardants to a convenient, house-plant size, no more than two feet tall, and bearing blooms no more than six inches across. Under cultivation, it is cut back twice a year.

The poinsettia was introduced into Hawaii, where it now grows wild along the roadsides and is often used as a hedge plant. Many Hawaiian poinsettias grow to twelve or fifteen feet, and have a shaggy double flower of a deeper red than our familiar hothouse products. There, too, the rose and cream varieties that are commonly found in Mexico are of striking beauty.

The poinsettia is related to snow-on-the-mountain, long familiar to gardeners, and also to fiddler's spurge, which grows from Arizona to North Dakota and suggests a dwarf poinsettia; its upper leaves simulate petals of bright-red flowers, and the tiny petals are often tipped with green.

The poinsettia's leaves and stems, like those of all

spurges, contain a poisonous milky juice. Man has discovered that in many species of *Euphorbia* this juice is of great economic value—castor oil produced from an African *Euphorbia*, rubber from a tropical American *Euphorbia*, and tung oil from the tung tree, a Chinese plant of the spurge family. But the poinsettia is truly the *Poinsettia pulcherrima*, and its gay crimson flowers are so long-lasting that it has become the favorite winter flower in millions of American homes.

POPPY

The poppies that sprang up on the battlefields of Flanders after World War I stirred many poignant memories. For thousands who saw them blowing between the acres of low white crosses they were the heart's blood of the young men who had died there. It was like an echo of the Flemish myth that declared that scarlet poppies sprang up wherever martyrs of the Spanish Inquisition died. In any event, their appearance seemed miraculous. The fields had lain barren during four years of war, the earth packed hard, trampled by thousands of soldiers' boots, or deeply scarred with trenches and shell holes. Then, suddenly, they were ablaze, the endless graves decorated with crimson poppies no one had planted.

But this was nature's way of healing the land and men's hearts. Poppies had grown in Flanders' fields since grain was first cultivated there. Their dramatic blooming after the war, which demonstrated how long the seeds lie dormant but remain viable, aroused the interest of botanists and horticulturists, who flocked to the scene. Actual soil counts were made, and it was found that more than 2,500 poppy seeds per square foot had waited out the war years.

The poppy is a prolific plant, a single one producing

10,000 to 60,000 seeds. Such fertility permits them to grow wild on the borders of fields and among the grain crops, where scabiosa and camomile daisies have also grown for centuries, survivors from a simpler era.

History

The scarlet field poppy, *Papaver rhoeas,* is a native of the Mediterranean area. It was grown in Egypt, where it was used medicinally and as a condiment, as early as 1500 B.C. In Greece it was a popular love charm, but Hippocrates, too, is said to have employed it as a medicinal. The Greek-born Galen, who became court physician to Marcus Aurelius, noted toward the end of his life that poppy seeds were good seasoning for "bread, comfits, and junketing." But the early Romans had used it for enchantments, and to cure the black melancholy that afflicts a lover.

It was with the Roman legions that the poppy spread across Europe, springing up wherever grain fields were planted to provide fodder for the horses. It moved northward from the Mediterranean in this fashion, its minute seeds always mingled with seed grain, through the Rhineland to Flanders, and so across the Channel to England.

One tradition holds that *Papaver,* the Latin word for the genus to which the poppy belongs, derives from a word the Celts used ironically, to imitate the sound their Roman overlords made when they chewed poppy seeds as a confection. According to another theory, the pap, or pulverized food, fed to infants was often doctored with poppy juice to induce sleep. Old wives' tales of the Middle Ages record that dried petals of field poppies were used to calm children who were suffering from colic or whooping cough; and they were grown as a medicinal in early monastery gardens. The Elizabethans, we learn from Gerard, believed that the leaves and seed capsules, boiled in syrup, induced sleep and relieved "pain, rheumes, and catarrhes." The field poppy, then, seems to have been endowed with some of the Oriental poppy's reputation, although in fact it contains no opium.

Whether its benign effects were real or imagined, it was also used to soothe fretful children in the early days of the American colonies.

Field Poppy in America

The scarlet poppy came unbidden with the first grains planted in the colonies. But it was also brought by many women for their dooryard gardens. It grew in New England gardens, and in the first gardens at Williamsburg, when that Virginia capital was laid out in the seventeenth century.

Botanical Bits

The field poppy is an annual that needs the sun to thrive, and is best grown in sandy loam, but it will adapt to most soils. Its average height is eighteen inches. Its dark-green leaves are irregular on a wiry stem, and its four scarlet petals are tightly folded inside a two-leafed calyx, which drops off as the bud expands, releasing the flower. Normally it is flame red, though very occasionally it may be white or purple; and all shades have a purple-black center. The roots are weak and thin, which means the poppy cannot be transplanted, but the plant itself has vigor and endurance. These characteristics have enabled hybridists to develop large numbers of ornamental poppies in a considerable variety of colors. Among the best known of the hybrids is the Shirley poppy.

Shirley Poppy

One morning in August 1880, the Reverend William Wilkes, vicar of Shirley, England, walked down his garden path to view his flowers and vegetables. It was a gray flagstone path, bordered with rows of bright red field poppies, one of his favorites. His attention was drawn to a corner of the back garden that joined the hayfield, in which poppies also grew profusely—according to many of the Reverend Mr. Wilkes's parishioners, a sign of poor farming.

He noticed that one of his poppies was quite different from any he had seen before; it had a wide white edge. He marked it and saved the seed, which he planted in a separate garden bed. It produced 200 plants the next summer; but the blossoms of only four of them had white edges. Challenged and curious, Wilkes saved their seed, from which the following year poppies of a dozen varying shades bloomed. Some were edged with white, some were striped red and white, and some were a pale red or pink; one was pure white. Strangest of all in these sports was the absence, or near absence, of the black cross in the center, characteristic of the field poppy. This was an exciting breakthrough. He now began breeding out the black centers entirely, and the results were astounding. Eventually he produced single pale poppies in every shade of pink, rose, salmon, and scarlet, with plain and ruffled edges, with white pistils and stamens, and even with a white base.

He named this new triumph the "Shirley poppy," for the parish to which he was vicar. It has proved a finer annual than the field poppy, and is as easily grown. He was a great rural naturalist, and it was not his only triumph. Among the others are the Shirley foxglove, with six-foot upright stalks and long racemes of huge flowers, and the Wilkes apple, delicious both eaten out of hand and cooked, which is one of the most popular in England.

The field poppy has many family rivals besides the Shirley, each providing color and beauty for the garden bed: the Oriental poppy, *P. orientale*, the largest and most ornate of the perennials; the California poppy, *Eschscholzia californica*, latter-day strains of which range from pale gold, through rose-pink and rose-orange, to carmine and crimson-scarlet; and the Iceland poppy, *P. nudicaule*, a perennial that asks little but sun and offers beauty in great flower cups of orange, apricot, cream, or rose.

But this is the story of the field or corn poppy; simple and enduring, it has been a symbol of rural beauty and tradition for thousands of years. John Ruskin sums up our admiration for this immigrant poppy, calling it the most transparent and delicate of all the blossoms of the field.

POTATO

High in the Andes, the village of Sorocota stood empty when the starving desperate conquistadors burst into its compound. The Indians had fled before the approach of the Spaniards, but they had left their campfires and their potatoes. Thus chronicled Juan Castellano, a member of the band and the first recorder of the white potato.

There is little doubt that the white potato is native to the Andes, and that it was completely unknown in North and Central America before the Spanish came to the New World. Alexander von Humboldt, the German naturalist from whom the Humboldt Current derives its name, traveled in Spanish America for five years. (See Scarlet Sage.) He cited the absence of the potato from North and Central America as evidence that the Indian civilizations in those areas were completely unaware of the Inca civilization to the south.

The Incas' agricultural practices far surpassed those in Europe. Indeed, nowhere else in the sixteenth-century world had crop rotation, planned planting, irrigation, and controlled production reached such a degree of sophistication. As many as 200 years before Pizarro's arrival, Andean Indians had even learned to preserve their potatoes by freezing them in glacial drifts, by drying them in the sun, and by storing them in cool caves.

Pizarro's soldiers ravished the Inca nobles of their gold and jewels. The padres who came with them coveted the rare fruit and vegetables with as much rapacity, and it is said that every treasure ship returning from the New World carried boxes of seeds and bulbs and plants. The abundance of the Andes invaded Spain, and a green revolution began in the monastery gardens, whose supervisors, the trained botanists of the day, recognized at once the splendid qualities of the *batata*. It grew quickly in a variety of soils, produced abundantly, and kept fresh in its corklike skin, so that it could be carried aboard ship to feed sailors, and exported to Spanish colonists—a factor of growing importance. It made its official entry into what is now the United States as an essential food shipped to the first Spanish colony in Florida, founded at Saint Augustine in 1565.

Europe and the Potato Invasion

Francis Drake, turning back to England in 1568 after some of his most triumphant raids on the Spanish Main, concluded by plundering the Florida coast. He then picked up the English survivors of Walter Raleigh's doomed colony on Roanoke Island and carried them home—with a supply of white potatoes, which they had bartered from the Indians, who in turn had probably obtained them from the Spanish colonies to the south. The potatoes are said to have been planted on Raleigh's estate in Ireland; thus began the history of the Irish potato.

Raleigh, an enthusiastic experimental gardener who brought many new plants home from his naval adventures, soon became a champion of the potato. It adapted readily to the Irish climate, and he believed it could be as valuable there as in the New World. He carried some potatoes and their flowering vines to London and presented them to Queen Elizabeth—an event to which Gerard refers in his *Herball*, remarking that inasmuch as the plant had come from Roanoke Island it could be called the "Virginia potato." His picture of it was the first ever published. But for another century, despite this temporary flurry of interest, the English paid scant attention to potatoes.

They did not fare much better on the Continent, to which they found their way clandestinely through the efforts of Charles de Lécluse (Clusius) and two members of the Fugger family. These three men, defying Spanish restrictions, made an audacious, secret excursion into Spain in 1576, to see for themselves the plants brought back from far places. They discovered several hundred new species, many from the New World, among them the potato. Lécluse, already famous through Europe as a collector of foreign plants, carried the potato out of Spain and introduced it into Flanders. Later he gave it special mention in his famous herbal, *Rariorum Plantarum Historia*. But Europeans accepted it only as a curious foreign herb, or, at best, a new decorative green: an exotic, best grown in botanical gardens.

The Irish Potato Colonizes Bermuda

In the course of the next fifty years, word about the potato's virtues gradually seeped out of Ireland. The story went that one and a half acres planted to potatoes could feed a poor Irish farmer's family of six for a year, and took very little labor compared with any other food crop. By the middle of the seventeenth century, England was suffering various side effects from the Reformation, and was in a state of great economic unrest. The potato sounded like a solution to the intensifying problem of chronic hunger, and the Royal Society of London, meeting in 1663, took steps to encourage its cultivation in England.

Following the lead of the successful Spanish colonial policies, England ordered the potato included in ship lists for all her colonies, as a "veritable stay for stomachs of slaves, servants and swine." Yet, with the shining exception of Bermuda, there is little record of the potato's having been grown in colonial gardens.

Supposed to have been discovered by the Spanish navigator Juan de Bermúdez in 1515, the islands remained uninhabited until a fleet of ships carrying colonists to Jamestown, under George Somers, was wrecked there in 1609. In the many months before the survivors were found, Somers fell in love with the islands and took possession of them for England. He investigated their cultural possibilities, and when his group was rescued he returned to England for supplies, determined to make the enchanted place a permanent English colony.

Somers, a member of the London Company and an enthusiastic promoter of American colonization, was familiar with Raleigh's "Irish root," and believed that it should do well in the rich soil of the Bermudas, despite the limited water supply.

Although most of the imported English crops did not thrive there, the potato went happily to work, produced abundantly, and saved the colony from starvation. How vital the potato crop was to Bermuda's very existence is clear from the early court records, which contain numerous accounts of potato stealing and the dire punishments dealt out for the crime. Potatoes soon became the chief cash crop of Bermuda, and were used in lieu of money, as tobacco was in Virginia.

The potato is said to have been sent from Bermuda to Virginia in 1621. Perhaps John Rolfe, who later married Pocahontas, was instrumental in the importation. He had been among the original settlers shipwrecked in Bermuda, where he had lost his first wife and child, and he had seen the potato produce abundantly and feed the colonists when other crops failed. He introduced into Virginia the superior West Indies tobacco that made the colony rich and famous. It seems likely, then, that he tried to benefit it by also proposing the growing of "Bermuda potatoes." But Jamestown tobacco had quickly become a success on the European market, and every inch of cleared land was planted to the "money crop." So the potato was neglected and forgotten for another century.

W. R. Van Dersal, discussing the origin of vegetables in *The American Land,* is one of many authorities who insist that the white potato was unknown in the mainland colonies until 1719, when it was brought in by Irish immigrants and grown in Londonderry, New Hampshire. He adds that for the next thirty years the Irish were the only settlers in the area who imported and cultivated potatoes.

Scotland, too, remained adamant on the subject of "Irish" potato culture. In 1728, to relieve a famine, England ordered the Highlanders to plant potatoes, but the Scottish government replied that "good Presbyterians would plant none of the heathenish food recommended by papist neighbors."

Potatoes and a French King

France accepted the potato even more reluctantly. It was not until early in the reign of Louis XVI, which the French Revolution would terminate, that Frenchmen took the *pomme-de-terre* to their hearts. His majesty, persuaded by Antoine-Auguste Parmentier—a distinguished agronomist who is sometimes credited with having invented what we all now call "French fries"—that the potato could do much to calm the restiveness of the hungry peasants, decided on an all-out campaign for its propagation.

He invited court aristocrats, nobles from key agricultural provinces, and foreign diplomats to a widely proclaimed all-potato dinner. (Among the foreign personages present was none other than Benjamin Franklin, an interested and able gardener in his own right.) Queen Marie Antoinette appeared with a potato blossom in her hair, the king with one in his buttonhole, and all the dishes served at the dinner, from soup to dessert, had been concocted from potatoes.

As a sequel, the king chose a field on the outskirts of Paris that peasants would pass on their way to market, and had it planted to potatoes. As soon as they began to blossom he put a guard around the field. This aroused the curiosity and the covetous instincts of the peasants, who watched with envy and commented derisively. What exotic food for the royal table was being grown with such care while they had barely enough to eat?

When the potatoes were mature, the guards, so prominent by day, were withdrawn by night. Slyly laughing at the king's foolishness, the peasants crept in, night after night, and stole what they considered the royal potatoes. Secretly they tested and tasted them, and found them good. Being French, they devised marvelous new recipes for the earth apple, and soon it was being cultivated in thousands of gardens.

François Regnault, whose famous *Botanique* of French plants was published while all this was going on, describes the potato as the only good thing that ever came out of America.

The Potato War and Napoleon's Prize

Frederick the Great, not to be outdone by Louis, issued a royal decree that potatoes should be planted throughout Germany for food and fodder, and to stave off famine in case of crop failures. The potato produced so abundantly that it was soon recognized as a crop essential to the national economy of the Rhineland.

The War of the Bavarian Succession, a typical example of "cabinet warfare" that occurred in 1778 and 1779, has been called the "Potato War" because success depended on control of the potato crop, and the entire campaign was spent in provisioning the armies. Still, it may have stimulated Russia's interest in the potato. Having watched the rapid and amazing success of the potato in the Rhineland, Russians began experimenting and found that it produced abundantly in their colder climate. They began growing it on an ever larger scale, and it soon became a staple food crop in Russia; today Russia is the top producer of potatoes in Europe.

Napoleon, too, became a potato promoter. During the

Napoleonic Wars the British blockaded French harbors, and the ships from the West Indies could not deliver any sugar. This caused such distress in France that Napoleon offered 10,000 francs to anyone who could make sugar from a French-grown plant. The potato came through again: a young man produced sugar syrup from potato starch and acid. It proved so successful that the potato-syrup business continued after the wars all ended. Among the factories opened in other countries was one in New York, and potato syrup appears to have been a familiar product as late as 1861, when Mrs. Beeton's famous book was first published. She describes it as very sweet, and says it may be used as a substitute for honey.

At the same time potato starch was replacing wheat starch in Europe and America. Starch was an essential of the age, necessary in the eternal laundering of ruffles and ruffs worn by both men and women. One of the first patents issued by the new Patent Office of the United States was for commercial potato starch. But the inventive and thrifty Yankee soon learned to make his own starch from the potatoes in his own garden patch.

The Black Days of the Potato Rot

The potato was fast becoming an international crop when tragedy struck. In the autumn of 1845, just as Ireland's crop was ready for harvest, the new potatoes turned black and rotted rapidly in field after field.

This seemed unbelievable. In the course of more than a century the Irish had become so dependent on potatoes that they had gradually stopped growing other foods. Potatoes supplied winter fodder for their cattle and hogs, so they grew little grain, and they had almost abandoned their kitchen gardens. Ireland had become a one-crop nation.

And now, suddenly, the crop was gone. There was virtually nothing to eat in Ireland, for man or beast. The people became frightened. Thousands died of hunger, and many of those who ate the rotten potatoes in desperation died, too. Others, who lived through the wretched year, planted what good seed potatoes they could save, but next autumn the black rot struck again, and more Irish starved and died.

By the autumn of 1847 the crop was very small, because most of the good seed potatoes had been eaten. When the black rot struck for the third time, the starving people streamed from the doomed island. They poured into England and Scotland, crowding all the port cities, seeking steerage passage to the United States and Canada. For those who found it, the voyage was a tragic one. Immigrants still cooked their own food on shipboard, and what food they had was often spoiled. The ships were overcrowded, and it was the season of storms in the North Atlantic. The weak and the sick died during the crossing.

When the survivors arrived they found the ports crowded with thousands of earlier Irish immigrants, often penniless and hungry. There was not enough work for all; there were not enough houses to shelter them; there was no money to buy food in this strange land. Their plight drew the attention of newspapers and politicians, and in due course gave rise to new immigration laws for the United States— laws that would control conditions on ships that transported the immigrants and make some provision for their welfare after they arrived.

All this called further attention to the Irish potato itself. Its increased cultivation by American farmers stems from this period, and eventually potatoes became our most widely eaten daily national dish.

An International Product

The potatoes that came with the Irish immigrants were not like those that we know now. Up to the time of the black rot the potato had changed little from its Peruvian ancestor. It was not until 1861, when Luther Burbank developed a truly new variety, that the potato won world popularity.

Early in his career, Burbank had noticed a potato plant with a large, unusual seed ball, resembling a small green tomato. Out of curiosity he planted the seed from this one fruit. Seventeen plants resulted, and one of them produced better potatoes than any hitherto known. After many seasons of selection and hybridizing, he named it the "Burbank russet." It achieved enormous fame, and remains important today, more than a century after its first appearance.

During that century the potato's career has taken on steadily mounting importance economically and nutritionally.

In Holland it is highly respected. Dutch potato growers are licensed, and candidates for the license are given free training at national horticultural schools. Perhaps, then, it isn't surprising that of the 5 million tons of vegetables produced annually by the Dutch, 4 million tons are potatoes.

Scandinavia, France, and Russia have developed national potato dishes that make delicious and nutritional contributions to their daily diet. In the Far East the potato has pioneered in feeding the hungry masses. Missionaries had introduced it by 1800, and it adapted quickly, producing abundantly in the Temperature Zones and in higher elevations of the semitropics. In India it is accepted as a native food, but Tibet and China, where potatoes were introduced a century ago, have continued to look upon them as a foreign food, to be fallen back on only in times of crop failure and famine. The present regime is campaigning for their increased cultivation. Japan, surprisingly, has so far neglected a vegetable well adapted to her climate; but eventually the modern potato will surely play a part in solving one of her great national problems—food for a growing population.

America grows 390 million bushels of potatoes a year, on 3 million acres of land, with Minnesota and Wisconsin ranking first and Maine and New Jersey not far behind. But 90%

of this, the world's most important food crop, is produced in Europe.

The Potato Goes to College

The United States Department of Agriculture has diligently investigated the life, health, and propagation of the potato. This has resulted in potatoes resistant to disease, and it has improved their flavor and cooking qualities, which depend on the starch content. A little starch means a dry, firm potato, best for boiling. A potato with a high starch content is best for baking, and makes mealy, fluffy mashed potatoes. But how are we to know what the bag on the grocer's shelf contains? There are more than eighty varieties of potatoes in use today, each with its special size, flavor, or cooking quality. Researchers at Cornell University have proposed that they be packaged and labeled as baking, mashing, boiling, or frying potatoes, instead of geographically as coming from Maine, Idaho, Long Island, or elsewhere.

So far, shoppers are still in the dark. That may be one reason why they have turned so eagerly to all the dehydrated, precooked, or frozen potatoes produced by modern technology. Of these, the dehydrated potato is the oldest. It was devised as early as 1901, but little was heard of it until the period between the two World Wars, when a large food company conceived the idea of a testing and tasting survey to be conducted among the married students of a large Midwestern university.

The company struck gold. The student wives responded to the product with the greatest enthusiasm, so its bulk manufacture began, and the potato shifted from the vegetable bin to the cupboard shelf.

What sort of personality will the most valued vegetable in the world offer us in the twenty-first century? The gods at the Department of Agriculture tell us that it will present a completely modern face, largely through new uses for auxins devised by plant researchers. The auxins are plant hormones, found in all members of the vegetable kingdom, that regulate plant growth. The natural auxins were first identified by F. W. Went in 1928; synthetic ones were developed two years later by P. W. Zimmerman. Plant scientists can now exercise almost complete control over plant culture. Man-made auxins are used to encourage more vigorous growth, promote flowering and fruiting and root formation, and produce seedless varieties. They can even increase a plant's resistance to cold, heat, or dryness.

With such power in the hands of botanical Merlins, the maturing of a potato crop can be controlled to adapt to climatic conditions, and the sprouting or deterioration of potatoes in storage will be controlled, so that they can be preserved for months longer. This means more and better potatoes, and has led to the establishment of the International Potato Center, in Lima, Peru, where agriculturists confer on publicizing the potato's superior food value and its world-wide cultivation to feed a growing population.

Sinister Aspects of the Potato

Why did the potato take so long to work out its destiny in a hungry world? There was in fact a sound basis for the suspicion that made Europeans resist it for so many years after its introduction. The potato, whose botanical name is *Solanum tuberosum*, belongs to the nightshade family, which includes the tomato and tobacco, other American plants that were invading Europe. But the potato plant closely resembles the European nightshade, which is a deadly poison, known as the "Devil's herb," and the potato's similarity rendered it suspect.

The potato does have a narcotic power, but it resides in the flower, the leaf, and the stem. During the seventeenth century, the more adventurous grew potatoes as a curiosity in their herb gardens, and some even tried them as fodder for swine. But a haze of mystery continued to hover over the

potato, which was said to possess potent aphrodisiac properties. Volumes of legend grew up about it, and even today
there are those who carry a slice of potato in their pocket or
handbag as a specific against rheumatism.

Yams, Sweet Potatoes, White Potatoes

The yam and the sweet potato were common in Europe a
century before the white potato, to which they are unrelated.
Confusingly, all three were originally called by the Tainan
Indians' word *batata,* and the white potato did not win its
distinguishing name—*patata* in Spanish, "potato" in English—
until late in the seventeenth century.

Perhaps much of the confusion can be attributed to the
earliest records, made largely from hearsay or by eyewitnesses
writing from memory. Few of these were trained observers,
none of them were scientists, and all of them were eager to tell
spellbinding tales about the wonders of the New World. In
these early records the white potato was often confused with
one of various plants called "groundnut," in this case *Apios
tuberosa,* which really belongs to the pea family but which
has tubers that are sometimes as large as three inches in
diameter. It is native to North America and an excellent food,
and it was commonly used by the Indians from our Atlantic
coast to the prairies. It is said the Dakota Sioux organized
large, efficient crews to harvest this "wild potato," as the
white man called it. E. M. Clute, in his *Useful Plants of the
World,* claims it was the groundnut that was carried back to
Raleigh's Irish estate from the Roanoke colony. And although
it is little known today, many botanists claim it is among the
very best of wild foods.

The sweet potato, *Ipomoea batatas,* is native to tropical
regions of the New World, and, as its botanical name shows,
belongs to the same family as the morning-glory (which see).
This immigrant from the Aztec civilization to sixteenth-century Europe was soon exported from there to the American
colonies. The yam belongs to yet another family, the

Dioscoreaceae; it is native to the warm regions of both hemi-spheres, and is grown commercially in both the Far East and the Americas. In our own country we frequently confuse it with the sweet potato, though there are connoisseurs who maintain that it is decidedly the more delicious of the two unrelated tubers.

Potatoes in Art and Design

During the period when the potato was regarded as a New World curiosity, Europeans and Asians alike were fascinated by its interesting leaves and flowers, and created dozens of embroidery and chinaware patterns from them. Among the most famous of these is a large pottery plate that was brought from France, long ago, to hang in the dining room of the old George Washington Hotel in Queen Anne, Maryland. Its subject is the promotion of the potato by Louis XVI, and it depicts a group of French cavaliers standing in admiring atti-tudes around a basket of potatoes. Its title is "Le Cercle de Pomme de Terre."

Recently the inroads of progress led to the closing of the quaint colonial inn's dining room. And so the plate, which had aroused the curiosity and admiration of diners for a century and more, was presented to Norman Taylor, a Queen Anne resident and one of the nation's foremost botanists.

Surely this is an appropriate ending for the plate entitled "Le Cercle de Pomme de Terre" and for our potato pot-pourri.

PRIVET

The early colonists were dismayed to find that privet grew nowhere on the Atlantic coast. In England it was an intimate part of their lives: it hedged their gardens and village greens, and it was essential to the varied work of farmer, fisherman, and housewife. Privet the colonists must have, and by general demand it was the first European shrub brought to the colonies.

The Need for Privet

Privet was a basic plant in rural Britain. Since Anglo-Saxon days its black berries had been used to dye homespun; its bark, in the process of tanning leather; and the hard white wood of its thin stems, for shoe pegs.

From Cornwall to Portugal fishermen had dyed their nets bright red with privet bark, because centuries of tradition said such nets brought in the largest catches. But privet was even more important to the farmers; before the days of manufactured rope and wire its long willowy branches provided the bindings for sheaves of grain and bales of hay, and fastenings for bundles and bags. It is from this use that Linnaeus chose the privet's Latin botanical name, *Ligustrum*, a binding.

Privet had always grown in Europe's monastery gardens. The juice of its black fruit had provided ink for cloister records and rural magistrates, and it was still used in the seventeenth century in the occasional letters exchanged by country dwellers.

Privet had also been a part of the physic gardens of cottage and castle because it provided a varied supply of medicinals. A powder pounded from the dried leaves healed ulcers, festers, scaldings, and burns; a decoction made from the berries was used as a mouthwash; an oil pressed from the flowers softened "constricted sinews," a common complaint of farm workers. Clusius, in 1601, was recommending that crushed privet blossoms "be laid to the head to swage the pain" of headache.

Topiary Triumph

Privet was first used as a decorative shrub in the gardens of the Pharaohs, when horticulture was considered a royal occupation. Centuries before the Christian era it hedged palace gardens in Egypt and Persia, edged their sanded walks, and circled their beds of bright flowers. Its tolerance to shearing and shaping made it ideal for topiary, and it was clipped and cut into triangles, obelisks, geometric shapes, and even into the fanciful forms of birds and animals. The Romans copied the practice and carried it to even greater lengths, clipping privet, boxwood, and cedar into elaborate representations of scenes on land or sea. Privet was particularly satisfactory for such purposes because it was native to the Mediterranean, easily propagated from cuttings, faster growing than box, and adaptable to the most radical pruning.

Lost during the Middle Ages, topiary art was reborn in the Renaissance gardens of France and England, and transported to our colonial south, where privet still reigned in formal gardens long after the designers of European gardens had turned to natural landscaping. (See Box.)

For three centuries, only European privet grew in American gardens, large and small. Then, early in the nineteenth century, Japanese privet was introduced to the West Coast. It spread rapidly, and began its triumphant trip across the United States as California privet. Today it is the most widely grown shrub in America, and the European privet is seldom grown except by nostalgic gardeners, although in the eastern states it lives on here and there as a wildling.

California privet, *Ligustrum ovalifolium*, is the fastest growing of the species. Its lustrous leaves and long lithe limbs produce a notably beautiful and compact bush, which has even more endurance to shearing than the European privet *L. vulgare*. Among other privets popular in American gardens is the ibolium, or ibota, privet, *L. ibolium*, and the dwarf ibolium, variety *regelianium*, much used along highways for binding the soil on banks. Ibolium is the hardiest of the privets and perhaps the most beautiful, because its horizontal branches bear pendant white tassels of flowers each spring.

The variegated privet, *L. japonicum variegatum*, with its green-and-yellow leaves, is dramatic when used as an individual specimen plant.

The privets belong to the Oleaceae, or olive, family, and are closely related to the lilacs. As Gerard noted: "The blue lilac and the white lilac lately come are like unto the common English Prim or Privet whereof doubtless it is a kinde."

Privet in Our Brave New World

Privet is tolerant of urban smog and the unnatural deep shade produced by high-rise buildings. Like people, it is adapting to modern conditions. Nowhere is this more evident than on Long Island, where miles of low clipped privet hedges once bordered the estates and the lovely gardens distantly viewed from the quiet roads that wound their way through Nassau and Suffolk counties. Today the intense traffic of the highways and the high cost of gardeners have made it almost mandatory to let the privet go its own way; and it has responded by producing thick green hedges twenty to thirty feet high, which absorb the noise, assure privacy for homes, and bless the motorist with fragrance and beauty when it comes into bloom each spring.

In today's gardens, large and small, the privet hedge provides an enduring green fence, a secluded enclosure, and an effective foil for rows of bright flowers and ornamental plants. It helps to keep the earth green, and continues to serve man's ever-changing needs.

PRUNE

There were no prunes in America when a French ship sailed into San Francisco Bay in 1849. The harbor was full of ships, but they were strangely silent and deserted. The reason was gold. Only a few miles away, streams and rocks were teeming with gold for the taking.

The excitement mounted aboard the French ship as each crew member began dreaming of instant wealth. And when morning came, the captain found himself alone.

Among those who had jumped ship was a young apprentice seaman named Louis Pellier, a farm boy from the south of France. Pellier did not strike it rich, but for the next year he wandered from one gold camp to another, panning what he could, and falling in love with the beautiful countryside. In the end he traded his small bag of gold dust for a few acres of land, and settled down near San Jose to farm. He knew that food would soon be more important than gold to the thousands of fortune hunters who were pouring into California; and by 1853 he had saved enough money to send to France for his brother Pierre, to share his future and fortune.

Three years later the Pellier farm and nursery were so well established that Pierre returned to France for his bride. He was also commissioned to bring back some nursery stock, particularly prunes, because it was the brothers' ambition to establish prune orchards like those that generations of Pelliers had cultivated at Ville Neuve d'Agen.

The accomplishment of the dream took some doing. First the saplings and scions had to be stuck into raw potatoes and packed in wet sawdust. After the long wagon trip to Bordeaux, Pierre and his bride faced repeated delays in sailing, and finally learned that no ship would go by way of Cape Horn to the west coast of California that summer. In desperation they took a freighter bound for Panama.

When they arrived in Colón, after weeks of sharing their quarters and their water with the saplings, they discovered that no ship sailing for the West Coast had room for a bride and "green truck." But their problem was solved by the completion of the new Panama Railroad, and the Pelliers and their prunes were among the first to cross the isthmus by rail. After another prolonged delay in Panama City they took ship for

San Francisco, and finally arrived in San Jose in December 1856.

They celebrated Christmas by planting the first prune orchard in California.

California Prunes

The French prunes flourished prodigiously in the California climate. They were propagated by cuttings and seeds, and soon a good-sized orchard was established and bearing. A third brother now joined Louis and Pierre, and they began the specialized work of curing prunes for commercial shipping. By 1863 they had won a responsive and growing market. But such success did not go unnoticed. By 1880 there were a number of large-scale planters with 100 acres or more of prunes, among them a German immigrant named Kamp, who was a trained orchardist, and who introduced many new and effective ideas.

Between 1880 and 1890 there was a phenomenal increase in prune production in the Santa Clara valley. Average trees were producing 150 to 300 pounds annually. By 1900, only thirty years after the first California prunes reached the commercial markets, a million pounds were sold. In production and shipping, California already outstripped France, hitherto the world leader in the growing and marketing of prunes.

This great good fortune inspired further emulation, and prune orchards were successfully established in Oregon, Washington, and Idaho. Prune orchardists in France, Spain, and southern Germany looked on with envy, and jealously guarded their European markets, while the real competition to California prunes came from pioneer orchards planted in Australia and South Africa.

Prunes and Plums

The Latin word for plum is *pruna;* thus the genus name, *Prunus.* Tradition holds that the plum, which may have been cultivated longer than any other fruit except the apple, was introduced into Greece by Alexander the Great, returning from triumphs in Syria and Persia. Plums are native to the North Temperate Zone around the world, and there were wild plums in North America when the white man arrived. There are in fact thirty species indigenous to our continent, and at least two of them deserve passing mention. The Indians made good use of the wild red plum, *Prunus americana*, which they ate both raw and cooked, and dried into nothing less than a sort of prune, which was a staple of their diet. Wild-plum butter is made from the red plum to this day. Then there is the low, shrubby beach plum, *P. maritima*, which adds its own charm to many of our eastern coastal areas, especially on Cape Cod, where beach-plum preserves are made in commercial quantities.

But these are trifling matters compared with the cultivated plums, most of which are derived from European and Japanese plums. Some excellent strains, however, have been hybridized from native species, and have the virtue of thriving in climates where plums with foreign forebears will not grow.

As for the prune, it is a very special kind of plum or *Prunus domestica*, notably meaty, of high sugar content, and pre-eminently adaptable to drying. Prunes had been introduced to the eastern United States in 1853, but the climate proved unfavorable and the orchards never developed.

Commercial Curing of Prunes

The first commercial prune crops were sun-dried. Insects, weather, and lack of visual appeal tempted orchardists to use caustics such as lye to hasten the drying and check the deterioration of the skins. They also tried glucose, indigo, and glycerine. But both the pure-food laws and progressive-minded orchardists brought an end to these crude practices. By 1930 dehydration tunnels, fired with gas and oil, had been designed. They dried the prunes to uniform perfection, and opened the doors for mass production. Today the average prune crop ranges around 170,000 tons, a record demonstrating the honorable place the prune enjoys among the immigrant plants that have enriched America.

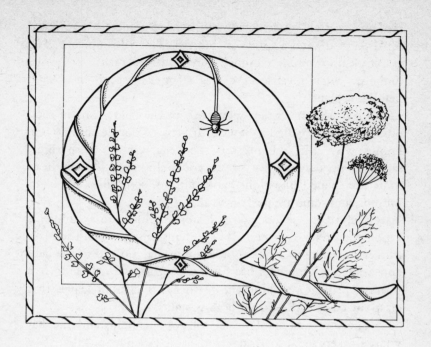

QUEEN ANNE'S LACE

Cool, white, and intricate, Queen Anne's lace hedges our summer highways with spendthrift loveliness. But who of us can imagine the full beauty and wonder of this plant as we speed by? Not even those who gather a bouquet of it can immediately grasp the complex structure of its blossom. It has dazzled many nature writers, who have diligently counted its florets and marveled at the genius of nature.

Here is the essence of a July afternoon spent corroborating the vital statistics of this amazing plant: The average flower head of Queen Anne's lace measures three inches across. It is a compound umbel, consisting of seventy-five smaller umbels that constitute a colony of flowers containing 2,500 tiny florets. Each floret is one-eighth of an inch in diameter and is itself a perfect flower, with five petals, five fused sepals, a pistil, and a spray of stamens. This adds up to 12,500 petals on one complete umbel. The plants average four to five umbels, which means there is a staggering total of 62,500 petals per plant.

All of the florets are white except the center one in each umbel, which is called a "jewel." It is royal purple, and no one knows why. But the mystery deepens. Of all these thousands of petals, few are symmetrical. In every flower head some are shaped like right-hand mittens, some like left-hand mittens, and some are perfect ovals.

Each of the seventy-five clusters of florets in an umbel is supported on its own stem. These stems radiate from the main flower stalk like the spokes of an umbrella. To add to the lacy appearance, each umbel has fourteen sepals, with three to five fernlike prongs matching the feathery, finely cut leaves that clasp the two-to-three-foot hollow stalks.

There is a hum of insects above the field, and the flat white flowers seem to glisten in the sun. Science says that they act as semaphores for the insects: the sunlight reflects from the thin layer of nectar in each tiny floret, and as the flower nods in the breeze it flashes messages of light and shadow. When the bees, butterflies, and other insects with long proboscises see these flashings, they know that the exposed nectar on the umbels is spread too thinly for them to gather, and they fly on to richer fields and long-necked flowers. But the flies and wasps, and the ants and other crawling insects, interpret the bright reflections as an invitation to a banquet. In return for the nectar served up for them on the shallow flower plates, they spread the pollen from floret to floret as they creep over the surface of the umbel.

A Nest of Seeds

Now that the plant has been pollinated, the tiny incipient seeds need more protection than the flat flower could give. At this critical moment, as if by magic, nature automatically closes each flower head, and the spokes of the lacy parasol draw toward the center, forming a "bird's nest," inside which the seeds ripen in safety.

In the hot August sun the bird's nest grows crisp and brown. It is said to be used by some types of bees for a nest; ants take refuge in it from summer showers; and I find many lined with silken webs of minute spiders who have sought temporary homes. Primarily, however, it is a storehouse for seeds. There they stand, row on row of seeds—2,000 in most mature heads; 10,000 to an average plant—arranged on their circular shelves around the walls of this concave umbel and held in place by their adjacent bristly hairs. What infinite wisdom engineered this efficient storage!

The seeds themselves are almost unbelievable in their complex structure and their efficiency. Shaped like a minute Japanese lantern, each one has five slender lacquered ribs—the space between being etched with hairs—and each seedcase bears two seeds and a supply of oil. While the seeds ripen, which takes two to three weeks, the whole umbel is saturated with a volatile oil which discourages birds and mice from eating the seeds and leaves a field of Queen Anne's lace smelling faintly like turpentine.

When the flower head is completely dry the seeds are ready for dispersal. Now the withered spokes release their hold, the storage shelves disintegrate, the brown sere umbel nods in the wind, and the minute seeds sail over the fields and roadsides. As they drop to the ground, the bristly hairs catch on weeds, grasses, and soil that will protect them until the next spring, when they will again provide a wayside garden for your delight.

All this infinite detail and accurate planning that one weed may live, be fruitful, and multiply!

How It Got Its Name

The history of almost every flower involves a great deal of human history as well.

When Queen Anne arrived in England from Denmark as the bride of James I, the wild carrot was still a novelty in the royal gardens, as we know from Parkinson, who reported in his herbal that at summer fetes the ladies of Queen Elizabeth's court had worn wild-carrot leaves instead of plumes in their hair.

Queen Anne often sat with her ladies in the summer garden at Somerset House. This garden, which she had leased in 1604 to John Gerard, possessed a border of wild carrots, probably planted by Gerard himself. Legend says that Anne challenged her ladies to a contest to see who could produce a pattern of lace as lovely as the flower of the carrot. The ladies knew that no one could rival the queen's handiwork, so it was a long summer ordeal for them and a triumph for Anne. Though the lace she made that summer has long since disappeared, the memory of her accomplishment lives on in the exquisite flower called "Queen Anne's lace."

Career of a Carrot

Queen Anne's lace, or wild carrot, is closely related to the garden carrot, which has been brought to its modern perfection only in the twentieth century. In fact they were once the same vegetable and they still bear the same botanical name, *Daucus carota*, which is a corruption of a Greek word and a Latin word, both meaning "orange."

The carrot originated in the eastern Mediterranean and was used by the early Greeks and Romans as a medicinal with a wide variety of attributes. Dioscorides says that it protects one from the "venom of wild beasts." Pliny the Elder suggests that it was used as a love potion, guaranteed to be effective, and Galen goes so far as to claim that it actually "procures lust." As a vegetable, however, the carrot remained a bitter, tough taproot, edible only in the early spring as a pot herb.

The first improvement came when it reached Holland in the fifteenth century. The Dutch, during the long wars of the

Middle Ages, had been often forced to substitute vegetables for meat, a trial from which a nation of masterly horticulturists emerged. They improved the size, flavor, and color of the carrot, and shipped it across the Channel to England with the boatloads of fruit and vegetables sent to help feed London's growing populace.

The carrot was first grown in England as an oddity in castle gardens, and admired for its flowers and fernlike leaves; but it proved adaptable, and the need for food soon led to its being grown in cottage gardens, from which it found its way into medicine chests as well as stewpots. Both Gerard and Culpeper recommend the carrot for numerous ills. Culpeper says that the carrot is influenced by Mercury, the god of wind, and that a tea made from the dried leaves should dispel wind from the bowels and relieve dropsy, kidney stones, and women's complaints.

The Carrot in America

The carrot arrived in America with the first settlers, but it must have been its easy adaptability and supposed medicinal qualitites that recommended it, because it was still a tough, stringy, greenish taproot that needed hours of cooking. And it still looked much as Queen Anne's lace does today. So it remained for nearly three centuries.

In America food crops grew prodigiously in the virgin ground, and in a few generations the carrot had been degraded to stock feed. It is still cultivated for this purpose in Germany, where it has also served as a source of sugar, and as a coffee substitute.

The dawning of the twentieth century coincided with a new interest in plant and food chemistry, a field to which attention increasingly turned during World War I. The attendant world-wide emergency in the production and distribution of food forced research in nutrition and the development of plants best qualified to protect the general health.

Among the vegetables that came out of obscurity was the carrot. When plant scientists discovered that carotene, an excellent source of vitamin A, was present in its leaves and roots, it was rapidly modernized. A tender, nutritious, appealing, and tasty orange carrot was developed for a vitamin-conscious generation. As a result, carrot consumption has tripled in the last fifty years. Today the per-capita consumption is fifteen pounds annually.

How did the pale and stringy wild carrot change so rapidly to the plump orange vegetable? Research revealed that carrots grown in temperatures below 50° F. are pale and thin, but that temperatures from 60° to 70° F. help to develop a plumper root, and increase the activity of the carotene, to produce a bright orange carrot with a lot of visual appeal. The carrot is now grown in the United States in the north in summer, in the south in winter, and in California over a longer period. Today's rapid transportation makes it available in our markets every month in the year, frozen, canned, or fresh. In Europe, incidentally, carrots appear in many shades of yellow, orange, red, or white, according to where and how they are grown.

Carrot Controversy

There seems to be a contemporary revolt against these scientifically perfect vegetables. A recent article in a national magazine featured the wild carrot as a delicious vegetable free for the gathering on any roadside, but cautioned that it is a biennial and should be dug in its first spring, as it is neither tender nor edible after it has bloomed. Cultivated carrots, of course, are grown as annuals and not permitted to go to seed.

Ironically, the carrot that the Puritans imported as a precious herb is now considered a pernicious weed from coast to coast, and North Dakota is the only state Queen Anne's lace has not invaded. Few weeds are so hated by farmers, but the agriculture experts recommend it as an excellent green manure if it is plowed under before it flowers.

Return to the Flower Garden

With the development of the cultivated carrot came a distinct division between the garden carrot and the wild carrot.

In England Queen Anne's lace was long considered a desirable plant for the flower garden. Kew Gardens grew it as late as 1891, and chose it as one of the first flowers offered to other botanical gardens. It was honored in early America, too. Old records say it grew in the Governor's Palace garden at Williamsburg, where it must have looked lovely. Planted in beds of dark-red coxcomb and pink globe amaranth, it was much admired for its white lacy flowers and ferny green leaves. But tastes change, and the wild carrot's tendency to spread banished it from our gardens two centuries ago.

Its return is the design of modern hybridists, who have created from it an exquisite new plant that looks like a great airy bouquet. Perhaps because it is listed in most catalogues under its old Latin name, *Daucus carota*, it still seems little known. But once again, in a few well-planned gardens Queen Anne's lace grows elite and lovely among the tiger lilies, offering various delights to all who pass.

QUINCE

We moved into an old house in a Pennsylvania valley in late spring. There were two bushes in the garden covered with exquisite roselike blossoms; one was white, the other pink. Their beauty, their strange name, and the romance that hung about their planting instilled in a child a lifelong curiosity about quinces.

The land had been one of the last possessions of the original Indians. The owner was a descendant of the daughter of an Indian chief, who had married one of the first Quaker settlers in the region. Through the centuries his great tracts of land had slipped away, until all he owned was this property, and a great grove of American chestnut trees beyond the town.

He had planted the garden of the old house with many varieties of pears, apples, peaches, and grapes, and with the two quinces, perhaps thinking to retire there and enjoy their bounty. But just at this moment the chestnut-tree blight arrived from Japan, and descended on his great grove of beautiful old trees. Unbelieving, he went to look at them, the remnants of a virgin forest; some were 100 feet tall and three feet in diameter. He felt a deep and personal kinship with them, and there they stood, dead and dying. Like his people, these native American trees had been attacked and decimated by an irresistible foreign force. His body was found dangling from a limb of one of the dying trees, and his garden, beautiful in flowers, fruits, and quince bushes, was left for strangers to enjoy.

History

Few fruits have changed so little through the ages. Before recorded history quinces grew wild in Kashmir, and had traveled all the way to Cydonia on the island of Crete. When the Greeks carried them to Athens, they assumed the name of the port from which they came—Malus Cydonia, "apples of Cydonia."

Through the Greeks they reached the Italian peninsula and became common in Rome, where they were held in high repute, and preserved in honey, used as a flavoring, and made into wine. They were also used as a love charm, and, says Pliny the Elder, they were reputed to protect against the Evil Eye, and were painted on the walls and eaves of Roman houses—much as hex signs would be painted on the barns and homes of early Pennsylvania Dutch settlers. During the long, sporadic excavation of Pompeii, which began in 1748, quince designs were found on many of the walls re-emerging from the ancient ashes.

Roman legions carried the quince to Gaul, where it grew

to perfection and won the hearts of the natives. The earliest accounts of making quinces into preserves say that the Gauls placed small branches, complete with leaves and fruit, in great vessels of honey and sweet wine, which were boiled over open fires until the whole was reduced by one-half.

Throughout the Middle Ages quinces were considered a choice fruit in France, and were espaliered on garden walls and grown in orchards with apples and pears. Rich sauces, preserves, beverages, and flavorings were developed from quinces in southern France, and at banquets they were placed among the edible fruits in great bowls, and are said to have exhaled an agreeable odor that was stimulating to the appetite.

It was in France that the ancient name "Cydonia" was corrupted to *cooin* and then to *quynes*, the plural of *quyne*— only a short step from the eventual English "quince."

As Christianity developed in the Germanic countries, quinces became a traditional holiday treat. Their bright-red candied slices were sold and served at Christmas festivals when few other condiments were available.

They spread to Spain and Portugal, and by the fifteenth century had crossed the English Channel. In Shakespeare's day they were planted widely in England, and used in preserves, as a sweetmeat, and for wine. In the seventeenth century Parkinson wrote that there was no fruit growing in the land that had so many excellent uses as the quince. Like other fruits in those days, it was considered a medicinal as well as a food, and was extolled as a "stomach's comforter."

In England it gained the reputation as a love charm that it had held in Rome. A box of quinces was considered a proper love token to be presented to a young lady as late as Victorian days. They are still popular in England, where the Portuguese quince is considered the best.

Quinces in the Age of Exploration

In the Age of Exploration the Portuguese led the way to Brazil, and around Africa to the Indian Ocean; and one of the

Spanish ships that had been under the command of the Portuguese navigator Ferdinand Magellan completed the first voyage around the world, after his death in the Philippines. These new adventures brought new problems. Scurvy plagued the sailing ships, which went so long without fresh food, and fruit that would keep was a necessity to balance a diet of salt-cured meat. For the officers of his ships Magellan chose casks of quinces preserved with raisins and honey—a delectable treat for the captain's table, but too extravagant for the crew. As his ships sailed on and on across the unknown Pacific, food gave out, scurvy ran rampant, and death took an increasingly heavy toll. It is a matter of record that those who survived ate the ship's rats and scraped away the inside of the empty quince casks, swallowing the sawdust for the last bit of nourishment that might have soaked into the wood.

Quince preserves were carried as an antiscorbutic, particularly for tropical expeditions, on Portuguese and Spanish ships. The Portuguese brought the first quinces to the New World and set out young bushes of them in Brazil, in orchards of apples and pears, a century before the Pilgrims reached New England.

Dooryard Quinces

The Pilgrims did list quinces as one of the essential and familiar fruits to be taken to New England. They were planted in the first gardens and orchards, and, like the early apple trees, produced abundantly after honeybees were imported to fertilize the blossoms.

Quinces were grown in dooryards throughout New England, and were carried west by pioneers. They lent their piquant taste to the perfect apple pie, or were sliced and cooked with sweet apples for a delicate company dessert. Crocks of quinces cooked down in honey, wine, or loaf sugar from the West Indies were kept in cellar cupboards and springhouses, and used in cakes, puddings, and meat dishes. In rural America quince-seed glue long served as an excellent adhesive, and the seeds soaked in small amounts of water were used as an aid to hairdressing and setting curls until the twentieth century.

Because raw quinces are hard and astringent, with a sandy texture, and become edible only after elaborate preparation, they have fallen into disrepute in this day of convenience foods and of the abundance, variety, and year-round availability of fresh fruit. By now there are only a few commercial quince orchards, mostly in western New York, but the quince serves the modern world in quite another way. It is used extensively as the rootstock for growing pears, for which there is a good reason. It is closely related to the pear, but has a much longer life and much more endurance—characteristics that it can transmit to the pear. Astonishingly, varied types of quince roots can also control the growth of the pear graft and determine whether the trees will be dwarf, medium, or large.

The success of the quince as an orchard rootstock has helped to establish American nurseries as our source of supply for orchards for 200 years. Before the Civil War, we had been dependent on Europe's nurseries. Since World War I, our American nurseries have become outstanding and our orchards independent of imported stock.

Botanical Bits

The quince, *Cydonia oblonga*, belongs to the rose family and is related to the apple and pear, which it resembles in fruit, leaf, and flower. It blossoms in late spring with exquisite and fragrant pink or white blossoms; its fruit should not be picked until after the first frost.

Quinces are adaptable to almost any soil, but appreciate light loam and humus. They may be propagated by suckers or layerings, are among the longest living trees, and produce a reliable crop each year. They are little affected by frost, demand little pruning, and are not damaged by birds, which avoid their hard fruit. Old books say that they need no spraying, but the quince beetle, or curculio, is now prevalent in orchards, and destroys the ripening fruit by eating into its heart.

Japanese Quince

The Japanese quince or japonica, *Chaenomeles lagenaria*, also made its way via Europe to American gardens in colonial days. It was blooming at Williamsburg before the Revolution and was recorded in other colonies by 1790.

Its durability and its prodigal blooms, in shades of dark red and pink and white, have maintained its popularity through the centuries. The ingenious French have long used Japanese quince bushes for enchanting hedges that flower fragrantly in May, stay richly green all summer, and in warm maritime areas provide fruit for preserving in the fall. Few ornamentals give so much for so little care.

RHUBARB

Rhubarb has played a part in the economic and dietary history of humankind for 5,000 years. What was it that compelled men to grow and use it for so long, in so many and such diverse circumstances?

The first rhubarb was probably introduced to America by the Russians, who were consolidating their fur trade in Alaska when the Founding Fathers were signing the Declaration of Independence.

Russia had worked hard for her American possessions. When Peter the Great persuaded the Danish explorer Vitus Bering to undertake an expedition beyond northeastern Siberia, there were no ships, no sailors, and not even any

shipbuilders on that remote coast. Contending with unbelievable hardships, Bering's men dragged lumber, rope, sails,
gear—and shipbuilders—across the Siberian wilderness on
dog sleds. That ordeal took two years, and it took another
year to build the ships, but in 1728 Bering sailed north
through the strait that now bears his name.

He found no new land on that voyage, and subsequent
expeditions met with little more success. Bering himself died
in 1741, after having been shipwrecked on the island that also
bears his name. But the survivors returned from that disas-
trous voyage with stores of fine sea-otter skins, which precipitated a rush of fur-seeking adventurers to the Aleutian
Islands. By degrees the fur trade was consolidated, although
the first permanent settlement—on Kodiak Island—was not
established until 1784.

The one great barrier against penetrating this Arctic
empire was scurvy. Few edible plants grew in those remote
and hostile parts, and they were suspect because the Eskimos
were reluctant to eat them, and avoided scurvy by eating the
stomach contents of ptarmigans and the organs of animals
they caught. But the expanding population of Russia's trading
posts needed a permanent solution to its dietary problems. It
was provided by two cold-climate plants introduced from
Siberia—rhubarb and cabbage.

Edible rhubarb, *Rheum rhaponticum*, thrived on the
tundra, a Finnish name the Russians gave the area between the
frozen sea and the forests. As they extended the fur trade
along the coasts and up the rivers, they took rhubarb with
them as an essential food and an antiscorbutic. Wherever its
strong roots could grow, it produced a crop.

But Russia's monopoly in Alaska met with strong competition from British and American interests, and in the early
nineteenth century her zeal for dominating the territory began to wane. Eventually, in 1867, she was glad to sell it to the
United States for $7,200,000 in gold. So Russian America be-

came Alaska—an Aleutian name chosen by Secretary of State William H. Seward, who had been virtually alone in his insistence on its purchase.

For a long time thereafter it was referred to as "Seward's Folly," but at last, in 1880, the discovery of gold in the Juneau region made Alaska a household word. Fortune seekers rushed to the strange unknown land, and rhubarb, which was one of the few plants they knew that was available, came into its own.

Today it is grown in many parts of the peninsula, where gardens flourish in the long hours of summer sunlight. In the Matanuska valley, famous for its rhubarb, it is cultivated by the acre outdoors, as well as in greenhouses: it has become synonymous with summer there. Fresh or canned, it provides an equivalent for fruit, and it is made into preserves and a potable wine. After two centuries, most Alaskans think of rhubarb as native to their state.

How Rhubarb Came to Europe

Early records show that medicinal rhubarb, *Rheum officinale*, a native of Asia, was used in China as early as 2700 B.C. For many centuries its dried and powdered roots, the source of a laxative long valued by the pharmacopoeia, were carried by camel caravan across Asia to the eastern Mediterranean and the Volga. In fact this may be how rhubarb got its name: Rha was the ancient name of the Volga, and "barb" is derived from the Latin word for the barbarian and unknown land beyond the Rha.

The medicinal rhubarb neither carried well nor kept well, and was therefore rare and costly in Europe, where it brought twice the price of opium, sixteen times that of myrrh, and far more than that of any spice. It was Marco Polo who decided to try growing rhubarb in Italy. He brought some from China packed in Chinese soil, in bamboo baskets, but after its year-long ride on camel back it languished under the hot Italian sun, and soon died. Europe continued to pay ransom prices for medicinal rhubarb.

As the sea route to India and China opened during the Age of Exploration, the markets of Europe were flooded with worthless and often dangerous drugs peddled by unscrupulous men. As a countermeasure, the University of Padua, the great medical school of Italy, opened the first public botanical garden in 1545, and established a chair of botany for a gifted and practical botanist named Prosper Alpinus.

In 1608 Alpinus decided to introduce Siberian rhubarb to Europe as a possible substitute for the exorbitantly costly Chinese rhubarb. He had some success, but knowing that rhubarb flourished best in colder places he sent seeds of it to John Parkinson and asked him to experiment with it. It appears to have flourished in the English climate, and in *Theatrum Botanicum*, Parkinson reported that it was being widely used as a pharmaceutical purgative. Siberian rhubarb had aroused the interest of other Englishmen, too. In 1630 Mathew Lister, physician to Charles I, who had accompanied a diplomatic mission to Russia, returned with a quantity of Siberian rhubarb seed he had acquired in Saint Petersburg. We may assume he planted it, with subsequent success. Culpeper in his 1649 herbal informs us that rhubarb is an excellent qualifier of unruly actions and the passions of Venus.

New Uses for Rhubarb

By the eighteenth century a lively trade had been opened with the Far East. Many Europeans discovered that various rhubarbs were grown in Asia and that they served purposes other than medicinal ones.

In the Himalayas, rhubarb was grown and used exclusively for what today's jargon would call "packaging." Its large leaves, often a yard in diameter, were a common sight in country markets, where vegetables and fruits were wrapped in them, or they were shaped into instant baskets for carrying other produce. Eastern bazaars sold a tasty and cooling drink made from edible rhubarb, and delicious and varied foods were created from the large unopened leaf buds, the flower buds, and the stalks. So at least it was reported.

These ideas spread slowly through Europe. In Flanders, France, and England, where edible rhubarb grew readily, people began planting it and cooking the stalks. Thomas Maeve reports in his *Universal Gardens and Botany*, published in 1778, that stalks of rhubarb were sometimes peeled and used in tarts. Nevertheless, rhubarb remained, on the whole, a curiosity.

Myatt's Rhubarb

In 1807 Thomas Myatt, who operated a large market garden in Lambeth, decided that rhubarb might have commercial value as a spring vegetable. He had received some roots from Flanders, where Dutch gardeners were growing it for sale, and one April morning he sent several bundles of stalks to the London market with his sons. It sold well, and there was a demand for more; his rhubarb soon became so popular throughout England that it was usually called "Myatt's rhubarb." He eventually devoted his whole garden to producing rhubarb, and continued to do so for the next fifty years. In gratitude for the prosperity rhubarb had brought him, he willed his garden to the city of London, and it became known as Myatt's Park.

Let us turn again to Mrs. Beeton and her marvelous 1861 compendium on household management. "Rhubarb," she says, "was comparatively little known till the last 20 or 30 years, but it is now cultivated in almost every British garden." She calls it "one of the most useful of all garden productions that are put into pies and puddings," and provides recipes not just for this duo, but for plain rhubarb jam, rhubarb and orange jam, and even rhubarb wine.

Rhubarb in America

Benjamin Franklin, traveling in Scotland in 1772, discovered the delights of a dish of fresh rhubarb and sent some seeds of the plant to John Bartram, whose botanical garden, the first in the colonies, was then in the making. In his note of thanks, Bartram writes: "Nov. 11, 1772. I have sowed the Rhubarb seeds in two places . . . in the sun and in a shady cool place. That which was in the shade grew . . . the leaf as large as my palm." Later on Peter Collinson, who sent Bartram plants and advice for thirty-seven years, included in one note a recipe for making rhubarb tart.

But rhubarb remained a foreign curiosity to most Americans until about 1800, when a garden-minded citizen from the state of Maine who was traveling abroad discovered the pleasures of rhubarb and decided that it would thrive in New England. He brought home some roots and shared them with other gardeners, who also became enthusiasts. Two years later Jonathan Lowell, in an account of this venture published in the *Massachusetts Agriculture Repository*, reported that rhubarb by then was generally available in New England markets.

By 1828 rhubarb roots were sold commercially by Thornburns, an early American nursery; David Landreth, the first American nurseryman to grow seeds for sale, was offering rhubarb by 1830. It finally became established as a dependable perennial in American gardens after the Civil War, and from then on provided that delectable dish of early spring, fresh rhubarb pie. And so it gained the familiar name "pie plant."

Botanical and Horticultural

Both garden rhubarb, *Rheum rhaponticum*, and Chinese rhubarb, *R. officinale*, belong to the Polygonaceae family of plants, and are relatives of dock and sorrel, and of buckwheat. Although there are a half-dozen kinds of garden rhubarb, the red varieties with grooved stalks and undivided leaves are the most popular. All rhubarbs are now propagated by root division, because seeds produce an amazing number of untried varieties.

Rhubarb is widely cultivated today throughout the North Temperate Zone. Ohio and Michigan lead in the production of commercial rhubarb, much of which is forced in greenhouses for early spring markets. It flourishes in Canada as well as Alaska, where it enjoys a long cold winter rest, but in the southwest it has reversed its life cycle to adapt to the warmer climate, resting during the long dry summer and growing in the cool winter months.

It has also adapted to the heat in southern Europe. Lacking a cold rest period, it produces great decorative leaves, and thrives as an ornamental or a potted plant, but it is never eaten there.

Rhubarb is an exotic plant; it is also suspect, because the leaves of every variety are poisonous. Only the underground roots of Chinese rhubarb are used medicinally, and only the succulent stalks of the garden rhubarb are now considered edible.

Although generations have used garden rhubarb as a spring tonic and mild laxative—with confidence and apparent success—no less a modern authority than the late Norman Taylor, who edited that useful volume *The Garden Dictionary*, tells us that this is a "completely erroneous notion" and that garden rhubarb contains no laxative properties at all.

RICE

One day in 1693 a battered trading ship that had long ago sailed from Madagascar accidentally found its way into Charleston Harbor in South Carolina. To pay for repairs and emergency supplies the captain gave the governor a bag of seed rice.

Charles Town, as it was first called, had been established as a stable colony only thirteen years earlier, after nearly a century of abortive attempts to settle the area. The new proprietary governor grasped at the bag of rice as an omen of fortune, and ordered some to be sown in his garden, as an experiment. When it seemed to thrive, the remainder was sown on the wetlands outside the town, where it flourished dramatically, producing in almost miraculous abundance.

Carolina's wide coastal plains, saturated by innumerable streams, indeed proved ideal for rice cultivation, and by 1700 it had become Carolina's leading export. More and more slaves were brought to the colony to work in the rice fields, and before long far outnumbered the white landowners and planters. By 1740 rice had made Charleston an international port and the fourth-largest city in America, a position it maintained until the Civil War, when the end of slavery temporarily disrupted Carolina's rice production. But rice dominated the economy of the state: in one way and another the labor problem was solved, and rice production continued to mount as the century moved to its close. By 1900 it had increased 83%, and almost 80,000 acres of Carolina's land were devoted to its cultivation.

Then disaster came. Rice stopped thriving, and within six years the 80,000 acres had dwindled to 19,000.

Research soon showed that man's greed and ignorance had brought about this abrupt change of fortune. On the mountain slopes above the coastal plains there were 20,000 square miles of virgin forest, where the streams that watered the rice fields had their source. This forest had been almost untouched, but in 1900 heavy lumbering began. As the trees were felled and the slopes laid bare, mountain springs failed and the streams silted up. And without an abundant, natural supply of water rice could not flourish. The lumber industry grew by 50% in one decade; in the same period, rice produc-

tion dwindled by 75% as Carolina farmers turned to potatoes
and tobacco.

The cultivation of rice spread rapidly to other parts of
the south, however, and in a few years Texas and Louisiana
were far outstripping Carolina. Today the cultivation of rice
along the Gulf Coast has become largely mechanized. Fields
are plowed by tractors, planted from airplanes, and irrigated
by canals between the fields, which at harvest time are
drained and allowed to dry out. Then the rice is cut by huge
combines that thresh and bag the grain and bundle the straw,
right on the spot. Comparable methods are employed in
northern California, where vast acreages of rice are now
grown.

Upland rice is cultivated in parts of the south, but the
yield from the high, dry fields is relatively small.

As for wild rice, a native of North America, it is obtained from a different grass plant, *Zizania aquatica*. The Indians harvested it in the northern lake region long before the
white man came, in a manner which tradition says involved
both sexes. The women waded through the marshy areas,
picking the stalks and beating the grain from them into the
canoes, which their menfolk paddled alongside. Later, men
and women joined in dancing on the dried grain to loosen the
hulls, and winnowed it by tossing it in the air, as part of the
harvest festival.

Wild rice is now gathered and winnowed by prosaic
modern methods and sold at premium prices in fancy-food
stores and in supermarkets.

Rice in the Orient

Common rice, *Oryza sativa*, is probably the oldest cultivated
grain in the world. Chinese tradition says that the Emperor
Shen Nung taught his people the art of rice cultivation in
2800 B.C.—an event that has been celebrated down the centuries in rural China on Rice Day, February 5.

In the Asiatic countries rice is interwoven with all the

basic needs of life. The words for "rice" and "food" are identical in several Oriental languages, and in truth it is the essential food, to which meat and condiments are but garnishes. It also provides alcoholic beverages, such as samshu in China and sake in Japan. From its straw the Asiatics make sandals and the wide-brimmed hats that protect them from sun and rain, mats for their floors, and thatch for their roofs. The hulls provide the fuel for the fires over which they boil their rice, and the brown layer beneath the hulls provides food for their animals. This must be removed before storing, because white rice keeps better than brown rice, just as white flour keeps better than whole wheat.

Rice is used as a motif in the most ancient Eastern art and plays a role in festival, theater, dance, and religion. Reverence for it is retained in the common daily greeting "Have you had your rice today?" Even our own custom of throwing rice at the bride is descended from the ancient Chinese fertility rite of pelting the bridal couple with rice to assure many children. In Japan, the roads through the rice country are lined with shrines to the god Inari, the rice bearer, and the rice goddess is prayed to in many homes.

To a great extent, methods of cultivating and harvesting rice in the Orient remain primitive. The sowing, transplanting, and weeding under water are all done by hand. Even the irrigation of the rice fields is often done with hand-turned water wheels, and at harvest the cutting, threshing, hulling, and bundling of straw are done by hand, too. By degrees, rice mills and mechanized irrigation are appearing in the Asian countries.

Rice in Europe

Rice was imported from India to the lands around the Mediterranean long before it was cultivated in Europe. It reached Greece as early as 300 B.C., and the word itself is derived from the Greek *oriza*, which meant "Orient." As rice became common in Europe, this was corrupted into *riso* in Italy, *riz* in France, *reis* in Germanic countries, and "rice" in English.

Let us look yet again to Mrs. Beeton, to find how rice fared in Victorian England. She remarks that it was held in great esteem by the ancients, who considered it beneficial for the chest and recommended it in cases of consumption. She herself praises rice, "a very valuable and cheap addition to our farinaceous food," and provides a dozen or more recipes for it. We learn from her that "of the varieties of rice brought to our market, that from Bengal is . . . of a coarse reddish-brown cast, but peculiarly sweet and large-grained. . . . *Patna* rice is more esteemed in Europe, and is of very superior quality; it is small-grained, rather long and wiry, and is remarkably white. The *Carolina* rice is considered the best, and is likewise the dearest in London."

Botanical Bits

Rice is an annual cereal grass native to the delta of the Ganges, in India, and probably to the deltas of other great Asian rivers. It has spread world-wide and is adapted to subtropical climates. It will produce two crops a year where moisture and warmth are available.

A single rice seed sends up many shoots, and the transplanted seedlings must be kept under water until harvest. The stalks grow from two to six feet tall, and their seeds are borne in a dense head on separate short stems similar to those of oats and barley. Rice is lacking in gluten, and is therefore unsuitable for breadmaking. Although it is inferior to wheat in nutrition, it is a light, wholesome food, and, supplemented by soybean products, to supply the essential protein, sustains life in the Orient today as it has since antiquity.

There are three main types of rice: Asian, Rangoon, and Carolina. Of these, Carolina rice is still internationally recognized as superior, as it was in Isabella Beeton's day—another immigrant plant that has enriched America.

ROSE

Cortez was welcomed to the New World with a festival of roses, as we learn from his fellow conquistador, the chronicler Bernal Díaz del Castillo, who relates that the Indians came out to greet these strange white gods bearing flowering branches and wreaths of roses. They threw chaplets of roses about the neck of Cortez's charger; they hung garlands of roses about Cortez's helmet; they strewed his path with roses.

Díaz also tells us that the road to Tenochtitlán was lined with grapevines, matted with honeysuckle and aloes, and entwined with wild roses, and he describes the gardens they passed as being perfumed with roses and flowering shrubs. When they neared the capital, Montezuma himself came out to greet them, wearing a sky-blue robe and carrying an armload of red roses.

Though some of this vivid picture may be attributed to flights of fancy, there is no reason to doubt Díaz's identification of the flowers as roses. The Spaniards were well acquainted with them—not only those native to Europe, but also the more spectacular ones the Moors had brought to Spain and cultivated in their walled gardens.

Champlain and Roses

No roses greeted Champlain as he sailed down the frozen Saint Lawrence a century later, nor did he record any wild roses growing in New France when he founded Quebec in 1608. But he was a zealous gardener, and when he returned from a visit to France in 1613, he brought back with him a bundle of roses to lay out a rose garden about his wilderness manse. One wonders what roses they were. A random guess is that he brought the *Rosa gallica*, the old French rose, and there must have been the damask rose, taken to France by the

Crusaders, and surely the eglantine brier and the yellow moss rose. Whatever they were, Champlain's roses are the first officially recorded as having been imported to America.

Roses in the Colonies

The Puritans reported in many letters and diaries that the wild meadow rose dotting the shores of New England and dunes of Cape Cod was a welcoming sight and a consolation their first springs. But they longed for the eglantine, the sweetbrier of their English homes and gardens, and soon began to import it. Kittery, Maine, claims the first sweetbrier from England was planted there; but it arrived on many ships, and quickly became naturalized, despite the harsher climate, throughout New England.

Adriaen Van der Donck lists the red carnelian rose and the white eglantine rose as having been imported from Holland in 1630, and as thriving in the walled gardens among the espaliered trees of New Netherland's first settlers. Some thirty years later Peter Stuyvesant boasted that his gardens held the greatest collection of native and foreign roses in the New World, some brought by Dutch East India Company ships from the far corners of the earth.

William Penn and his Quakers thought of English roses, too, when they laid out their City of Brotherly Love—perhaps being slow to discover the wild roses that grew so abundantly along the Delaware and the Schuylkill. Penn, who notes in his diary that he has ordered eighteen English roses for his estate at Pennsbury, believed in their medicinal and food value, as we discover in a *Book of Physic* that he wrote for his settlers. It is well seeded with information about rose cures and the use of roses in cookery. One wonders if a recipe entitled "To Comfort Ye Brains" was a cure for homesickness.

In the southern colonies every plantation had its rose garden, often patterned on those of England and France, and supplied with the rarest roses from the West Indies and

Europe. George Washington himself hybridized a new American rose from the *Rosa satigua,* and it was grown and sold in the south for many years.

A Wedding and Roses

The marriage of the Portuguese princess Catherine of Braganza to Charles II of England had a far-reaching effect on the story of the rose.

By 1660, Portugal had been trading in the East for nearly two centuries, and both the Portuguese and the Dutch had fought to keep England from gaining a foothold there. But Catherine's dowry, which included entry to the ports of Bombay and Canton, opened a door to the Orient for England at last.

The very first East Indiaman that sailed into London's harbor carried two dormant rosebushes that had been wangled from a mandarin's garden by an English officer. They flourished in the moist English climate; the British marveled at their size and beauty, and demanded more—and more—and more!

Plants and seeds soon became of prime commercial importance in the China trade. The British East India Company opened botanical gardens in Singapore and Calcutta, trained the crews of its great sailing ships to care for the plants, and sent plant hunters out to explore the resistant East for new botanical wonders.

As China roses appeared in English gardens, America demanded them, too. European nurseries grew rich propagating and shipping them to the colonies, and gardens from Maine to Georgia reveled in their fragrance and beauty. And then a new era in roses began.

Hybridization

China had practiced hybridization since antiquity, but it was the work of individual gardeners, an expression of their creativity and skill, and never done for commerce.

When Europe discovered the technique of hybridizing, it became a mania. In 1789 a China rose was crossed with an old European rose. The result was perfection. Not only was the hybrid larger, more fragrant, and sturdier than either of its parents, but also it retained the exceptional blooming period of the China rose, from May to October, whereas the European rose bloomed only in autumn. The new hybrid caused a revolution in rose culture.

Private gardeners and nurseries competed in creating new roses to bring them fame and fortune, and the results enriched the gardens of America, too. The steady stream of hybrid roses arrived here from France and England with romantic names to match their beauty: "Coup d'amour," "Rien ne moi surpasse," "Belle sans flatterie." The old moss and shrub roses, and the crimson damask of country gardens, gave way to the subtle fragrance and elegance of the first hybrid roses. But these, too, had their day. The first wave of hybridizing reached its height by 1850, and then the hybrid teas were born.

The hybrid tea roses are a comparatively recent development. Although J. B. Guilot, the French rosarian, was shipping hybrid teas to America by 1867, and the London nursery of Paul & Sons in the 1870's, it was not until 1890 that the hybrids were officially accepted as truly innovative and their tealike fragrance recognized as a distinct characteristic. Since then they have maintained a high place in American gardens, with few rivals in rich color, size, and aromatic scent.

The tea roses ruled through the Age of Innocence, a few quiet decades during which pleasant gardens and front porches were a part of summer. And then the perpetual hybrids were born of the Bourbon roses. These were even lovelier, longer-blooming, and larger than the original hybrid teas. Hybridization again became the madness of the moment, but now science and the laboratory took over. Simple experiments by the amateur were smiled at, and the search for new roses abroad lost its impetus.

Roses of the Past

Roses had been cultivated for more than 3,000 years in the gardens of China, Persia, Egypt, and the Greek islands. In all these places they were revered and enjoyed, honored in temples and palaces alike.

Through the centuries they have been used as a dependable food and medicine. Hippocrates, Pliny the Elder, and Dioscorides all record rose cures for ailments of teeth, ear, stomach, lungs, and intestines. As a complement to meals roses reached their height at Roman banquets, for which shiploads of them were imported from Egypt. Swags of roses decorated the walls, the floors were strewn with roses, the diners' couches were perfumed with rose petals, petals floated in the wine cups, too, and each guest wore a crown of roses on his brow as an antidote for too much wine. In the Roman code of honor no gossip exchanged at the table, above which the swags of roses hung, could be repeated outside the banquet hall. Thus the phrase "sub rosa" has implied confidentiality or secrecy through the centuries.

Because of such associations, the rose became a symbol of wine, women, and indulgence—of Roman degeneracy—and the early Christians refused to allow it to be brought into the churches or used in association with the Virgin. But during the Middle Ages it was grown in monastery gardens, used as a food and medicine, and finally as a decoration. Roses were carved in wood and stone, often above the door of the confessional. Prayer beads, originally made from rosewood, became "rosaries." Later these were often created of a paste of ground rose petals and salt by the lady of the castle as well as in monasteries.

Roses in Stillrooms and Kitchens

There are hundreds, perhaps thousands, of rose recipes preserved from earlier times: every Oriental and European country has its own cherished collection. During the Middle Ages the rose garden was an asset to the manor house, and a rosebush was an important part of a dowry. In the stillroom the mistress dried, ground, preserved, and mixed rose petals and hips with camphor and musk or, more often, with herbs she had gathered. She distilled rose water, and made lotions, ointments, and the flavorings and vinegars so essential for seasoning meats, fish, cakes, and conserves. To read these old recipes in this day of instant foods and mechanized kitchens is to be dazzled by culinary initiative.

ROSE CONSERVE

Take thy rose petals and cast thereto sugar, a good porcion. Mulberries, dates, chopped for a time. Now cinnamon and powdered ginger. Seethe on a slow fire the space it takes to walk a mile or two. Add almond milke, or if thou wilt, sweete creme of Kyne. Seethe a bit and pour forth the mess into thy crockery pots. To store happily, cover with linen full soaked in bee's wax.

All the herbalists pay tribute to the rose. Culpeper sums up their confidence in its medicinal value by declaring that "Rose syrup, rose honey, rose aromaticum, rose water, wine, and oil, and ointment" are superior for "humoring human aches and ouchings."

American Roses

North America has twenty-six native species of roses, and they are found from the Atlantic to the Pacific, though principally in the grasslands of our central states. Ninety percent of the roses cultivated in America, however, are of foreign origin. Although they were initially unsuited to our soil and climate, Americans have insisted on growing them, and have ignored their own riches.

The only truly American rose to be widely cultivated is the prairie rose, *R. setigera*, a native of our Midwest that was introduced by Samuel Feast & Co. of Baltimore about 1836, and won popularity as a symbol of national expansion during the great westward movement. Its hybrids, marketed as Queen of the Prairies and Baltimore Belle, are our hardiest climbers, but little is heard of them today.

Other American natives that still win a degree of public recognition are the early meadow rose, the Carolina or swamp rose, the Arkansas rose, and the pasture rose, which look much as they did in yesteryear. The only exception is the Virginia rose, a hybrid of the pasture rose introduced to gardens as long ago as 1768, which strove to compete with early imports.

The famed Cherokee rose, once believed to be a dashing native of our plains, is an immigrant after all. André Michaux, the French plant hunter who spent fifteen years in America, identified it as one of the Oriental roses introduced by the early Spanish from the Moorish gardens of Spain. The Cherokee Indians, the birds, and the pioneers spread this lovely China rose through our west, where it eventually became naturalized.

From the early 1500's, throughout Spanish America, the choicest of cultivated roses were imported for the gardens of haciendas and monasteries. Few specific records have survived, but many legends of roses have been handed down, among them a version of the story of Adam and Eve, who were driven from the Garden, the padres told the Indians, for picking a rose. The Indians would not have understood the traditional version because apples were unknown in Spanish America, but the cultivated roses of the mission garden were a constant temptation to them.

As Spanish missions and ranches spread through the southwest, roses escaped from walled gardens and grew wild; a great profusion of them greeted Father Junípero Serra when he arrived in Monterey in 1770. He had plodded many weary miles through sand and sun, and he records the miracle of finding the rose of Castile thriving in the wilderness. He saw it as an auspicious omen for the chain of Spanish missions he envisioned founding in California. He ends his diary note in gratitude: "Blessed is He who created the rose."

America's First Perfumery

Among the earliest products of the Shaker colony at Lebanon, New York, was rose water, which became nationally and then internationally famous. The austerity of the Shakers' life was symbolized by the Rose Rule, which stipulated that no rose cut from their fifty acres of herbs and flowers was to be used for decoration or personal enjoyment. To ensure that no one indulged in such vanity all the roses were picked without stems. Discipline and roses proved highly profitable, and the Shakers were soon shipping their products abroad, to compete with the perfumes of the Old World.

The Queen of Roses

While France and all Europe were being torn apart by the Napoleonic Wars, the Empress Josephine collected roses. At her insistence, Napoleon appointed one Dr. Delile, director of the Montpellier botanical garden, to accompany the army as a plant collector on the Egyptian campaign. Delile brought back the *Rosa alba* and the *Rosa centifolia,* which aroused so much enthusiasm that Napoleon assigned plant collectors to future campaigns. Roses poured into France as a sequel to each of his military successes, and in her star-shaped garden at Malmaison Josephine collected 250 varieties. More important, she used her private funds to stimulate horticultural research and hybridizing. She discovered and subsidized the great botanical artist J. P. Redoute, whose folio *Les Roses,* an exquisite early example of colored prints, included pictures of 200 Malmaison roses. She supported André Michaux, who established an exceptional plant exchange between the United States and France. Finally, her roses became the inspiration for the most comprehensive rose gardens in the world.

L'Hay-les-Roses, now owned by the city of Paris, was created by Jules Gravereaux, founder of the Bon Marché stores. It includes a library of rose literature, and a group of gardens that present various chapters in the story of roses—wild, Oriental, hybrid, and national. There is also an experimental garden, and a garden created and maintained in loving dedication to Josephine, which includes all the varieties of roses she grew at Malmaison.

An Ancient Essence in the Modern Mart

Josephine's intense interest in roses through the turmoil of the wars helped sustain the art of perfumery in France, now centered in Grasse. It had been founded on the fragrant damask rose that the Crusaders had brought from the Holy Land, where they learned the technique of distilling its essence from the Arabs. The countries of the East had learned to steep roses and create scented waters and oils, and had burned rose leaves for incense before recorded history.

Today 85% of the essence called attar of roses is produced from damask roses in Bulgaria. It takes from 180 to 300 pounds of petals to make one ounce of rose oil, but the oil of the damask rose is so fragrant that a single drop is enough to perfume a gallon of cologne.

Iran is famous for damask roses, and in Syria they grow wild. Modern scientific research has demonstrated that the dry soil and hot sun of these areas greatly increase the volatility of the oil and the scent, and also cause the roses to produce a great number of petals—factors that mean financial success for growers and perfumers. Iran alone exports over 300 tons of dried rose petals annually.

Operation Rose

The wild dog rose of English lanes and roadsides was another of the long-neglected plants whose value came to light again when Britain was cut off from imports of fresh vegetables and drugs during World War II. Nutritionists discovered that dog-rose hips possessed more vitamin C than most fruits and vegetables, and England organized "Operation Rose" to gather and distribute hundreds of tons of rose hips annually. The government manufactured rose syrup and rose pills to prevent malnutrition, and in army hospitals rose hips proved to be an invaluable "green medicine" for bone cases and hastened the recovery of the wounded.

The dog rose was honored for its war work and given a place in the Ancient Order of Herbs of Grace.

America's Rose Dilemma

Roses are endemic to the Northern Hemisphere, and our continent is particularly rich in native ones; yet we ignore them, and fill our gardens with roses introduced from Europe and the Orient. *The Encyclopedia of Roses* notes that 421 varieties of roses are now propagated in the United States for commercial purposes. At last, gardeners are turning from this welter of hybrid beauty and demanding old-fashioned roses that America thinks of as her own.

There are nostalgic requests for the seven sisters rose, the old familiar climber, with its flat clusters of seven roses, ranging in color from red to white, which once grew over the doorways of many a log cabin. But this rose of pioneer days was brought from China in 1821. The old hedgehog rose, which is seldom found today except in corners of forgotten graveyards, was brought from Japan via France after the Civil War. Even the old American beauty was a renowned French rose. These, and the crimson ramblers, the old shrub roses, the *Rosa rugosa*, and the moss rose all arrived in our East Coast ports and were carried through the backwoods and sold by plant peddlers before the days of catalogues. For more than a century the roses poured in, and many of the so-called wild roses of our lanes and roadsides are immigrants escaped from early gardens. Even the famed salt roses of Cape Cod dunes floated ashore from some wrecked China clipper.

The interest in these early roses is growing, and many are

now being rediscovered, collected, identified, and offered for sale. Today, new roses are launched by the American Rose Society and the Federated Bureau of Plant Industry, whose supervision assures us the best of future roses, both native and introduced. Perhaps Fate has destined the untapped treasury of our true native roses for the peculiar delight of future gardeners, who will follow their wild sweet scent down the garden path.

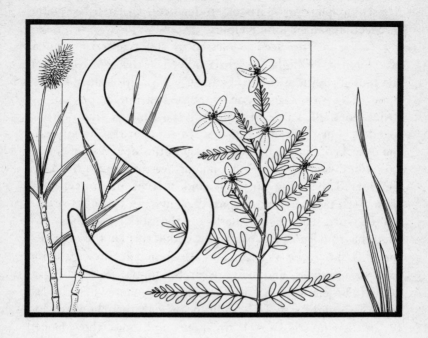

ST.-JOHN'S-WORT

No other plant has entered our country with the dramatic clamor that accompanied the first recorded arrival of St.-John's-wort.

In June 1696, a ship reached Philadelphia bearing a band of Rosicrucian pilgrims who were fleeing religious persecution in Germany. The Rosicrucians were a group of ardent fanatics of vague origin and tradition, accused of heresy and occult practices. In the weeks of passage they had remained suspect by other passengers and the crew because their only equipment for settling in the new land seemed to be a great telescope, an astrolabe, a sundial, and a supply of herbs believed to have magical properties.

There were forty men in the band, a motley bunch dressed in pilgrim's garb or student robes, led by a young intellectual named John Kelpius.

Their venture seemed blessed by supernatural guidance: after months of flight they had arrived in this safe harbor on the most propitious moment of the year, Midsummer Eve. A sense of gratitude and wonder at such a miraculous coincidence compelled them to observe the time as tradition demanded. They must not put their fortunes at risk by ignoring the powers of elusive divinities, even in this new world.

They looked to the highlands northwest of the city, shouldered their pilgrim packs, and hurried off to keep their tryst with mystery and magic, carrying the branches of St.-John's-wort, the sacred herb of the sun, that they had brought from Europe. Toward the end of the day they reached the hills above the Wissahickon River, where they gathered piles of branches, and wove their St.-John's-wort into garlands for their hair and necks. At the moment when the last rays of sunlight struck the hilltop, they lit their bonfires and began their mystic rites. With shouts and songs, they danced about the fire while the flames leaped high and higher. As the moon reached the zenith, they cast their garlands of St.-John's-wort into the fire, and when the incense of this enchanted herb rose among them they seized the largest of the burning logs and rolled them down the hillside—symbolic of the power of this herb to exorcise the Evil One.

Prayers, praise, and incantations were offered until dawn. But the fire burned on through Midsummer Day, until at last the rays of the setting sun again struck the hilltop, and the dying embers released the exhausted pilgrims from their ancient rite.

Kelpius and his band settled in Germantown, outside Philadelphia, where they laid out their herb garden with beds of healing herbs, cooking herbs, and herbs of worship. Beside the sacred St.-John's-wort grew wormwood, betony, monkshood, foxglove, and dill—old herbs with the highest potency for the casting of spells. In the midst of this garden, Kelpius established his sundial and his curious observatory, where he studied the stars and cast the horoscopes of each member of the community.

Kelpius's knowledge of plants soon brought him friendship with John Bartram, and, like Bartram, he corresponded and exchanged plants with Peter Collinson, the London merchant whose letters to American gardeners provide much information about early American horticulture.

Kelpius, whose fanatical self-discipline included fasting and undermined his health, died eight years after his arrival in Philadelphia; but his friend and disciple Dr. Christopher Dewitt, another brilliant intellectual and devoted gardener, assumed leadership of the Rosicrucian community, and subsequently opened one of the earliest botanic gardens in the colonies.

St.-John's-wort was also highly valued by the many religious sects of the Pennsylvania Dutch. Their name for it was the "blessed herb," and they prized it for the protection they believed it gave a newborn child. Some sects believed as the Rosicrucians did, in its power to discipline the Devil on Midsummer Eve, which they, too, observed. A traditional custom was to hang a sprig above the door to banish the Evil Eye, and at the same time placate the good fairies as a symbol recognizing their power to control evil with this herb.

The automobile has ended the isolation of these onetime church-centered villages, in which, until World War I, German continued to be spoken in many schools, churches, and homes. By now the folklore, rituals, and superstitions going back to the Middle Ages have withered away.

History and Romance

St.-John's-wort is thought to have originated in the eastern Mediterranean. Centuries before the Christian era it was recognized as an herb of destiny, marked by the sun god as his peculiar possession, and used in the sacrifices connected with the worship of the sun.

For such primitive faith there is ample explanation in the very structure of the plant. The leaves and golden flowers are clearly marked with tiny translucent circles through which the light shines, making them look like miniature suns. The flowers themselves, which grow in loose golden heads, seem like representations of the sun. The five rounded petals are dominated by long stamens that radiate from a five-styled pistil. Each stamen is capped with a tiny golden ball of pollen, so that every floret looks like a small sunburst. The natural habitat of this heliotropic herb is a sunny hillside, and it possesses what earlier generations considered the supernatural property of turning its face from east to west each day as it follows the sun across the sky.

As civilization pushed westward, the plant spread around the Mediterranean and became entangled with the myths and legends of Greece and Rome. *Hypericum*, the ancient lovely name for the genus to which it belongs, is derived from that of the Greek Titan Hyperion, father of Helios, the sun god, who drove his golden chariot across the heavens between dawn and dusk, and in Rome was later to be called Apollo. Hecate, the goddess of ghosts and sorcery, gathered *Hypericum* in the sunny fields of Greece. Circe, daughter of Helios, and Medea, princess of Colchis, distilled their potent charms for good and evil from its leaves and flowers.

As Rome absorbed Greek culture, the worship of the sun continued, and *Hypericum* was burned at the Festival of Fires, which observed that awesome moment of the summer solstice when the sun seemed to stand still. But as Roman civilization advanced, Midsummer Day became a largely secular holiday, a community thanksgiving for the sun's light and warmth in the past, and a plea to Helios to continue his blessings to earth. When darkness fell on Midsummer Eve, bonfires were lit in every Roman market square and village crossroads, and the people danced and sang around them and cast their garlands of *Hypericum* into the flames in lieu of ancient sacrifices.

Roman ships carried *Hypericum* to Britain, and Viking vessels to the northlands, and with it went the celebrations of the summer solstice, blending the traditions of Celts and Saxons with the rituals of the Druids.

The Christian Church and Devil's Flight

The early Christian church, unable to control ancient nature worship, suddenly discovered that Saint John the Baptist was born on Midsummer Day. So the Feast of Fires became the Feast of Saint John, and the pagan herb became St.-John's-wort. The word "wort" implied a medicinal herb, and the church declared it a plant sacred to Saint John, who had blessed it with healing power.

But the appeal of the supernatural is not so easily erased. The Crusaders brought back varieties of St.-John's-wort from the Holy Land, and with it came old tales of its properties in casting spells for good and evil. The Crusaders' St.-John's-wort flourished in the moist English climate, and with it flourished a renewed secret interest in old pagan rites.

In isolated villages throughout Britain, clandestine fires burned again on hilltops on Midsummer Eve. St.-John's-wort, monkshood, foxglove, and vervain were cast into the flames to placate the fairy folk, and as sacrificial incense in honor of the sun. The secrecy increased the enthusiasm of the devotees, who laughed, sang, and danced about the fires, leaping into and over the flames, intoxicated by their own frenzied exertions. The revels lasted until dawn, all moral discipline abandoned, and any child conceived on Midsummer Night bore no stigma, for it was said he had been fathered by the sun god.

St.-John's-Wort as a Medical Herb

As the power of the church increased, its discouragement of pagan rites became more effective. The monks collected all familiar herbs, recorded their ancient lore, grew them in their

352   GREEN IMMIGRANTS

physic gardens, tested their healing properties, and devised prescriptions for the sick and needy who flocked to them for healing. Thus St.-John's-wort won amnesty in monastery gardens. The monks revealed that its greatest property lay in the healing of wounds, and in the relief of inflammation of the lungs and throat. John Gerard wrote enthusiastically that St.-John's-wort was a most precious remedy for deep wounds, and he gives a recipe for a compound oil made of its leaves, flowers, and seeds, declaring that there was none better in the world. And the twentieth-century herbalist Florence Ransom suggests St.-John's-wort as a remedy for colds, coughs, and chest complaints.

St.-John's-Wort in American Gardens

Though St.-John's-wort does not appear in the earliest lists of herbs planted in New England, New Netherland, or Williamsburg, we may assume that many English and German settlers who came on those small crowded ships brought packets of its seed for their first gardens. We do know that before the Revolution it had escaped from these early herb gardens and spread along the lanes and roadsides from Maine to Spanish Florida.

Swedish botanist Peter Kalm found a form of *Hypericum* "about the lakes and rivers of Quebec," and John Bartram reported the discovery of a *Hypericum* in Georgia in 1776. But these, too, could have been runaways from the gardens of early European settlers.

During the eighteenth century new varieties of St.-John's-wort arrived from England—one of which was called "Aaron's beard" because the flowers had a multitude of extremely bushy stamens—and a large-flowered hybrid was imported from France. In 1862 *Hypericum patulum* arrived from Japan, and was followed by other Oriental species. Besides these, about two-dozen native varieties have been found, and a few of them have been disciplined by cultivation and are offered by nurserymen.

But St.-John's-wort is a social plant which, although it has been banished from man's gardens and may have left magic behind, has found a new form of service: it is now used by many states as an effective and beautiful ground cover for eroded highways banks where nothing else will grow.

SCARLET SAGE

Meeting in Paris in 1798, naturalists Alexander von Humboldt and Aimé Bonpland discovered a mutual frustrated desire to investigate the natural history of foreign countries.

Both young men had been attached to scientific expeditions that had been disrupted by the Napoleonic Wars. Now they joined forces and set off on a botanizing trip through the Atlas Mountains. But they were turned back by the sultan of Algiers, and resigned themselves to traveling through Spain. There, quite by chance, Charles IV granted them the rare privilege of pursuing their research in Spanish America. Thus began one of the world's most important expeditions, which led to a number of far-reaching scientific discoveries, and resulted in the importation of many new plants to Europe.

It was while they were traveling in New Andalusia, Mexico, that they came upon a field ablaze with color. It was dotted with a plant whose brilliant scarlet flowers were being pollinated by golden hummingbirds, and Bonpland declared himself in a state of ecstasy. Upon examination the flowers proved to be two-lipped, many of them an inch long, blooming in terminal clusters; the stem of the plant was square; its slightly fragrant oval leaves were toothed: it was obviously a member of the mint family, and therefore related to the European sages. Bonpland and Humboldt added specimens of it to their collection, and in due course it was classified as *Salvia splendens*.

Scarlet Sage in Europe

The shipment of plants, bulbs, and even of seed from their native wilds to commercial nurseries and botanical gardens was an ordeal for collectors because it often ended in heartbreaking waste. Bonpland's and Humboldt's collections were no exception; some specimens rotted while awaiting passage, and one shipment was lost at sea. But two great chests shipped from Trinidad under the care of John Fraser, a Scottish naturalist whom Humboldt had befriended in Havana, arrived intact. This collection, along with thirty-two chests of plants that accompanied Bonpland and Humboldt when they returned to Europe from their expedition in 1804, resulted in the introduction of forty new tropical American plants to Europe.

The cultivation of scarlet sage started soon thereafter in Germany, and we learn from a gardener's dictionary published in 1724 that a red sage from Mexico, *S. fulgens*, was also being grown on the Continent by then. But both of these seem to have been unknown in England until 1822, when *S. splendens* was brought directly from Mexico to London. *S. patens*, which bears intense blue flowers, arrived in 1838. Both were greeted with condescension by some English gardeners, who found their colors too gaudy among the cool pastel blossoms of their herbaceous borders. So it was that scarlet sage, which adapted so easily and spread so rapidly, found itself among the surplus nursery stock that was being shipped to the United States.

Scarlet Sage in America

Scarlet sage also rushed here directly from Mexico in the early 1840's, under the auspices of Joel Roberts Poinsett, our first ambassador to the Mexican Republic, and his friend William H. Prescott, whose great historical works *The Conquest of Mexico* and *The Conquest of Peru* helped to arouse widespread interest in the exotic plants that grew south of the border. (See Poinsettia.) Robert Buist, a Philadelphia seeds-man, forwarded the cause of scarlet sage, which gained fur-ther popularity in 1848 when Texas declared her indepen-dence from Mexico and pioneers discovered such wildflowers as bluebonnets, bitterweed, Indian paintbrush, and, in the hill country of West Texas, another scarlet sage, *S. roemeriana*, which shone among the blue Texas dayflowers.

During this century the popularity of scarlet sage has grown steadily in the United States, where it is treated as a tender annual. It will bloom from seed from early summer to fall, bearing brilliant flowers on plants ranging in height up to two feet or more. Its ease of propagation and its effective display make it popular in both home gardens and public plantings, factors that have encouraged nurseries to develop new hybrid strains. A few years ago the Welwyn salvias burst on the market with pink, salmon, mahogany, white, lilac, and purple varieties that have been remarkably true in type and color. Today's catalogues often include several pages of flowering salvias, from dwarf and compact border plants to three-foot giants, in pastels and in blazing colors. But *S. splendens* has no rival.

The American sages are closely related to the European salvias, of which 800 species are known. Most of the culinary sages come from Europe, where *S. officinalis* has been con-sidered a medicinal for numerous ills since the Middle Ages. *S. sclarea*, or clary, a handsome plant that once grew in the Vatican gardens, is still universally used in cooking, wine making, and perfume. And sage of one sort or another may well be the most popular seasoning herb in the United States: we import 2 million pounds of it each year.

SUGAR

Sugar is considered a staple in every modern household, but until the last century it was a rare and coveted luxury. Behind your carton of sugar lie 4,000 years of history, involving high adventure, mandarins and emperors, piracy and slavery, science and invention—and fabulous wealth.

The earliest records of sugar are in Sanskrit, the ancient Aryan language of India, to which sugar cane is native; it grew wild on the plains and was used by man before recorded history. But the conversion of cane juice was costly, and the transportation of sugar itself difficult and dangerous. Early Indian caravans carried bundles of sugar cane to China; even in this state it was considered a most desirable item, worth its weight in gold.

Later, Indian merchants introduced refined sugar to the Chinese court, and demanded an exorbitant price for it. Eventually a Chinese mission was sent to India to learn how to cultivate and refine sugar. The venture proved highly successful; within a few years sugar cane was thriving in southern China, refineries had been constructed, and the finished product was carried by caravan throughout the nation. Its use spread, ultimately, to Arabia and Persia, and then to the eastern Mediterranean.

Alexander the Great introduced sugar to Greece, as one of the prizes brought back from his invasions of Asia Minor, Egypt, and India. But in Europe it remained a rare, exotic food for another millennium.

Sugar and Crusaders
During the Crusades, from the eleventh to the fourteenth centuries, many Europeans spent prolonged periods in the Near

East, where they learned the delights of sugar. After their return they demanded its importation, and ships soon began pouring sugar into eastern Mediterranean ports, particularly Venice. But it remained a virtually priceless rarity because it was perishable, and expensive to transport.

To guard against loss, merchants ordered it wrapped in palm leaves at the mills in India, Arabia, and Persia, and then sewn into heavy cotton bags. These were sealed in chests stored in the driest holds in the ships, which carried red sails to identify them and were given priority in loading and unloading at ports. But pirates attacked the red-sailed ships, so sugar remained rare and costly, and was reserved for medicinal purposes, and as an adjunct to the feasts of the wealthy.

Sugar and Marco Polo

Setting out from Venice in 1271, Marco Polo traveled to China with his father and uncle, who had penetrated as far as Kublai Khan's eastern capital at Kaifeng in 1266. This new expedition by caravan reached Peking in 1275. Polo remained in China some seventeen years, as a privileged member of the Chinese court and a favorite of the khan. During his stay he visited many parts of the vast empire and studied its economics, including the cultivation and refining of sugar, which was so important to the merchants of Venice. He reached home again in 1295, and a year later was taken prisoner while participating in skirmishes between Venice and Genoa. In the famous account of his adventures that he dictated during his captivity, he called the sugar refineries the greatest marvel of China and one of the wonders of the world. Throughout the Renaissance his book remained for Europeans their chief source of information about the East—and about the manufacture of sugar.

By 1450 Venice had become the greatest sugar port on the Mediterranean, and tales of the adventures of the many Venetians whom it had made fabulously wealthy challenged the imagination. One ingenious Venetian discovered a way to change loose refined sugar into hard, cone-shaped loaves, which facilitated storing and shipping—a method that proved so successful it was used for more than 300 years. Throughout our own colonial period, sugar was sold by the loaf, which explains why so many mountains in early America were named "Sugar Loaf."

Sugar and International Rivalry

It became obvious to Spain and Portugal, watching Venice grow in wealth and power with the growing demand for sugar, that they must bow to Venetian dominance of the Mediterranean and seek new sea routes to the Indies, as well as new sources of sugar.

Portugal sent Gonçalo Velho Cabral out into the unknown about the year 1430, and he returned after having rediscovered the Azores, a group of uninhabited islands, fertile, warm, and only 800 miles off the coast of Portugal, which had been almost unknown since the days of the Phoenicians.

In payment of a heavy debt, the king of Portugal made a temporary grant of the islands to his aunt the duchess of Burgundy. The duchess, an ambitious woman, induced a group of Flemish farmers to emigrate to the Azores and develop plantations of sugar and oranges—a venture in which they were highly successful.

Columbus's first voyage was the Spanish answer to this challenge. His discovery of the West Indies led to the early planting of oranges and sugar cane in what is now Haiti. Both plants found an ideal home in the West Indies, and in a few years sugar was being exported to Europe in steadily increasing quantities. Eventually the Caribbean islands rivaled India as a leading sugar-producing area of the world.

Cortez carried sugar to Mexico, where it also found a happy home, and the remains of his original sugar plantation mills still stand near Cuernavaca. Its success in Mexico inspired the founding of great sugar plantations throughout the coastal areas of Spanish America. Wherever it was introduced the Indians were enslaved to work in the fields and mills, and when they either wouldn't or couldn't stand the forced labor, slaves were imported from Africa. Portugal followed the same course after having introduced sugar to Brazil.

Sugar remained the New World's most profitable export for more than two centuries and created a pattern of life: great wealth for a few, slavery for many. Because it represented large investments in land, labor, and transportation, it remained a luxury in Europe, and the common man continued to use honey for sweetening.

Sugar and New England Sea Captains

While sugar fortunes were being made in tropical America, hardy New Englanders were finding shipping more lucrative than farming. Yankee ships made the West Indies' produce an integral part of New England economy. Oranges, lemons, sugar, and molasses were in demand at all colonial ports on the northern run, and Yankee rum, fermented and distilled from West Indian molasses, a by-product of sugar, was readily sold on the southern run. The manufacture of rum became an important New England industry, and the increased demand for sugar created an increased demand for slaves. So the notorious Triangular Trade was born. New England ships traded Yankee rum in Africa for slaves, and slaves in the West Indies for molasses, which was made into more rum in New England. Thus sugar reached into many New England towns, and fortunes were made by the ships' captains, whose homes are still the show places of some old seaports.

Sugar Plantations in the United States

About 1750 Jesuit monks in Mexico sent sugar-cane cuttings to a monastery in New Orleans, believing that the Mississippi delta would be an ideal place for a plantation—as, indeed, it eventually proved to be. Through ignorance about sugar cultivation, this first venture failed. In 1780 sugar cane was again sent from Mexico to the New Orleans monastery, this time with slaves to do the planting and to teach the culture and refining of sugar to the monks. This cane flourished, and Spain proceeded to introduce sugar to other plantations along the Gulf Coast and eventually in Florida.

The young but expanding United States took note, and before long great sugar plantations were thriving along the southern Mississippi and its tributaries. New sugar barons established their fiefdoms, and the slavery that was essential to profitable sugar fields spread rapidly.

As competition for the rich sugar profits grew, new and more modern techniques were developed in crushing mills and refineries. Juice was extracted at high temperatures, and crystallizers spun off a rich green-black syrup called "black-strap" or "New Orleans molasses." This remained cheap, and the molasses pitcher became a fixture on the pioneer table and a part of western cooking. Refined white sugar remained a luxury.

But the luxury was humbled at last, brought low by one of the lowliest of vegetables, the beet.

Beet Sugar—a Man-made Rival

While our great sugar plantations were growing on the Mississippi delta, Napoleon was conquering Europe. When England finally blockaded French harbors, he offered a prize to anyone who could find a substitute for cane sugar that could be raised in France. It was won by a German scientist, Andreas Sigmund Margraff, who invented a process for extracting sugar from the common yellow beet, *Beta vulgaris*, which yielded from 2% to 5% sugar. Hybridization and selection eventually produced beets that yielded 12% to 15%. Napoleon then offered grants of land and cash to anyone who would construct mills to extract the sugar, and they proved so successful that within a few years beet sugar had largely replaced imported cane sugar in France.

The cultivation of sugar beets soon spread throughout northern Europe, and proved a real bonanza. Beet sugar sold at a fraction of the price of imported cane sugar; moreover, the green tops of the beets could be used as fodder for farm animals, and the molasses and residue of the beet mills was pressed into winter feed. It was also discovered that sugar beets enriched the soil, and that many crops grown in rotation with them flourished markedly.

Beet Sugar in the United States

American entrepreneurs took note of the rapid development of this new industry, and concluded that the prairies of the Middle West should be ideally adapted to it. Sugar beets were introduced to the north central states in short order, and produced abundantly. But they could not be processed in cane mills and refineries, and no one knew how to build and operate a refinery for beet sugar.

A mission dispatched to France bought a complete beet-sugar refinery, dismantled it, and shipped it to New Orleans. Steamboats carried it up the Mississippi to one of the great plantations, where it was met by a train of oxcarts and carried across the prairies to the beet fields. But the venture was a dismal failure because no one had experience in operating the machinery, and the sugar produced was bitter and inedible.

Eventually, experimentation and Yankee ingenuity triumphed, and in a few years beet sugar became available on our domestic market in great abundance and at a reasonable price.

Competition led to the rapid mechanization of cane fields and refineries, and reduced the price of cane sugar. In the early years of this century, sugar sold for five or ten cents a pound in the smaller grocery stores. Today, 3 million acres in

Florida, Louisiana, and Texas produce most of the cane sugar for domestic consumption. But sugar beets provide one-third of the world supply, with Russia, Germany, and the United States leading the field. Moreover, there is no perceptible difference in the end product: to all intents and purposes, beet sugar and cane sugar are identical.

Sorghum

Sorghum has long played an interesting role among American sweeteners. Although it is little known in the north, it is commonly used in the south and southwest, where it competes with New Orleans molasses, especially in rural areas.

Sorghum, which was introduced to the United States from its native Africa as a cheap substitute for sugar to feed the slaves of our south, is, like sugar, a member of the grass family, but it is not adapted to large-scale, mechanized production. Its popularity as a plant grown for sweetening has been confined to areas where small local mills produce molasses from it. Even today sorghum festivals are held in many rural communities, to harvest the crop and to operate the crude mills, which produce an estimated 32 million gallons annually. Sorghum molasses is high in food value, but this is only incidental to a sorghum devotee. And no wonder—for sorghum mixed with real butter is an ambrosial topping for hot biscuits!

Other varieties of sorghum provide an important forage crop. Of the sorghums grown on the Great Plains and in the southwest, about half are used for silage and half for feed grains. Broom sorghum, introduced 200 years ago, has been used in the manufacture of brooms and brushes ever since.

Sugar in Hawaii

The Polynesian voyagers who first settled the Hawaiian Islands brought with them coconuts, sweet potatoes, and sugar cane, which they called "ko" and planted about their dooryards. When Captain Cook arrived there in 1778 sugar was growing wild—a portent for the future, inasmuch as sugar would play a major role in transforming this lovely Eden of Polynesian primitives into a modern commercial society.

During the nineteenth century the Americans and Europeans who followed the missionaries to Hawaii grasped almost immediately the possibilities for Hawaiian sugar. The first mills were built in 1836 and the future looked golden. All that was lacking for unlimited success was labor. (See Pineapple.) The problem was ultimately solved—after the native Polynesians, and workers imported from Malaysia, Japan, Puerto Rico, and Central America had all been tried and found wanting—when John Whipple, scion of a missionary family, brought a shipload of contract labor from China, bound to serve from five to fifteen years in the cane fields. Thereafter sugar provided rich returns, and by 1920 it dominated the island's industrial life, produced 3% of the world's supply of raw sugar, and brought in $2.5 million. Since then, mechanization has ended the need for mass labor in the sugar industry.

But sugar—seconded by pineapples—has transformed the very physiognomy of the people, and an interracial society, in which native customs and character have been merged, has replaced the Polynesians who lived there two centuries ago. Now sugar itself is being replaced by tourism as the leading occupation and money-maker, and the results are devastating. High-rise apartments cut off vistas; highways cut through landscape; tunnels cut through mountains; and millions of transients have changed the face of the islands and driven away the old paradisaic leisure.

Sugar in Fiji

It was inevitable that Europeans would take example from Hawaii and attempt sugar cultivation elsewhere in the South Pacific. Great sugar plantations were begun on the Fiji Islands in 1877. Indentured workers were imported from India, and housed in isolated camps. Taken from the slums of India's cities, impoverished and starving, they throve in the sun and air of Fiji. They did not intermarry with the natives but their numbers increased rapidly, until today they represent 60% of the islands' population.

Otto Degener, a naturalist who has spent much time on Fiji, tells us that the native population is declining and is being replaced by the descendants of the sugar "slaves"—who, freed by the mechanization of the sugar industry, are now small shopkeepers, civil servants, and contract farmers for refineries. Sugar has changed the face of Fiji, too.

Sugar has also played its part in the Philippines, where it is one of the chief agricultural products, as it is in most tropical countries.

Explosive Use

The consumption of sugar has increased at an alarming rate in the last century. The average American consumes around two pounds a week, compared with half a pound in 1800, and the English average is 102 pounds a year per person, a fivefold increase. In the affluent countries almost all prepared foods, both canned and frozen, contain at least some sugar. Nutritionists and members of the medical profession now warn us of the perils of eating sugar, which is not an essential food in itself, despite the theory that it supplies quick energy.

Few plants have affected history more than sugar has, and we may doubt that its own history will wind down to an inglorious end. But we must recognize it for what it is, and use it with common sense and informed circumspection. Radical changes in the story of sugar may occur as our century draws to its close.

SWEET PEA

About 300 years ago an Italian priest, Father Franciscus Cupani, was assigned to the spiritual fields of Sicily. Nothing could have pleased him more, because he had a green thumb and an observing eye when it came to fields terrestrial.

As he wandered about the hills of Sicily he was attracted to an insignificant, weedy flower with a haunting fragrance. Indeed, its exquisite odor seemed its greatest asset. Its six-foot viny stem, which climbed over roadside weeds, its leaves and tendrils and small flowers, dull red with purple standards—all these made it look like an undernourished wild pea.

The Sicilians seemed to know little about the climbing weed with an irresistible perfume, but Cupani wrote a description of it that was published in the *Hortus Catholicus* in 1697. His article must have been widely read, for letters poured in asking for further information and even for seeds.

One of the men to whom Cupani sent seeds was Dr. Robert Uvedale, headmaster of a grammar school in Enfield, England. He was also a collector of rare plants from all over the world, and the undoubtedly affluent owner of six greenhouses—perhaps the first in England—in which to grow them. So important was his collection that when he died it was bought by Robert Walpole for his famed garden in Norfolk.

Under such patronage the Sicilian weed quickly came into its own in the cool damp climate of England. Hundreds of people traveled many miles to see it and marvel at its fragrance, and Uvedale generously shared the seeds. By 1722 sweet peas were well established, and Thomas Fairchilds, a nurseryman, recommended them for London gardens, describing them as "the sweet-scented pea of red and blue color." By 1730 they were advertised in Robert Furber's famed catalogue *The Twelve Months of Flowers*, which contained the first illustration of the plant ever published.

From the moment they were made available to the public their popularity grew, although their only real charm was fragrance. A century after their introduction they varied little from the wild peas Cupani had found, bearing two short-stemmed, dull-red-and-blue flowers. Only six color varieties had been observed by 1860, and the difference among these was negligible because sweet peas are self-pollinating and few crosses occur naturally. In spite of their fragrance they were being overshadowed by the flood of exquisite foreign flowers then being cultivated in Europe.

The Birth of Modern Sweet Peas

Trevor Clarke, an English botanist, decided to do something about reviving their popularity, and tried his hand at pollinating them. Forced hybridization was a new idea among nurserymen and one that was producing startling results with other flowers. And so it was with sweet peas: Clarke achieved a blue-and-white hybrid so attractive that it seduced other nurserymen into experimenting.

Henry Eckford, who became an outstanding sweet-pea breeder, began work in 1870, and in the course of thirty years totally transformed the dull wildflower. There were but fifteen varieties when he began. He developed 100 more in his Shropshire nursery, without sacrificing their precious fragrance. Sweet-pea fanciers from all parts of the world came to see and learn, and to participate in this botanical adventure.

One of them was Silas Cole, gardener for the Earl of Spencer, who some time later discovered among his own sweet peas a mutation from Eckford's Pink Prima Donna that was quite different from any other sweet peas known at the time. The whole flower was large and luxurious, with a small standard and waved and frilled petals. Further experiments proved it could produce a wide range of captivating colors, plus four blossoms to a stem. It had no odor but great charm, and proved to be only the first of an endless line of new varieties, each lovelier than the last. Cole named it the "Countess of Spencer" in honor of the earl's wife, and the "Countess" led the serenade that opened the sweet-pea era.

A bicentennial exhibition was held in 1900 at the Crystal Palace in Sydenham, to celebrate the 200th anniversary of the introduction of sweet peas to England. To the exhibition, at which 264 varieties of sweet peas were shown, came enthusiasts from every corner of Britain, including amateur gardeners with small home plots and professional gardeners from large estates. The excitement aroused led to the founding of the National Sweet Pea Society in 1901.

The London *Daily Mail* stepped forward to promote this fragrant fuss, and offered a prize of £1,000 for the best new hybrid sweet pea. Hundreds of gardeners took up the challenge, and dozens of new varieties were developed by professional gardeners, nurserymen, and amateurs. Out of all this came a real sensation, the first ruffled sweet peas, which were developed in three different nurseries at the same time. There were, of course, variations, because they were all the result of mutations; but it turned out that only one of them could bear seeds. This hybrid was an unqualified success—luxuriant, dainty, and prolific, blooming in a variety of colors from purest white to deep royal purple.

The Spencer hybrids were aptly named the "Grandiflora sweet peas." Their light, airy, varicolored blossoms looked like bright butterflies resting on a pea vine, because the standard, the large posterior petal, had now virtually disappeared, and the winged petals had been enlarged and become exquisitely colored.

The sweet pea was a favorite of Queen Alexandra, and became the flower of Edwardian England. Its exquisite fragrance and delicate beauty filled the silver epergnes at dinner parties, banquets, and weddings, and it was used in the extravagant corsages, boutonnières, and nosegays of the affluent and privileged—a symbol of an age of elegance and ease.

But a new age was opening, in which all fashions and customs would be commercialized for the masses, and the sweet pea bowed to the times. By now it is no longer a novelty. It is raised on a mass-production basis, shipped by air, advertised by city florists, and sold at relatively modest prices.

Sweet Peas in America

The sweet pea, *Lathyrus odoratus*, belongs to the pulse family and takes its name from the Greek word for pulse or pea. It is not native to North America, where it was unknown until the eighteenth century; but it was brought to the colonies on its first wave of popularity in England, and by 1750 was growing in the governor's garden at Williamsburg, where its delightful scent was much admired. Although it was still a European novelty, it was soon cultivated for its fragrance in many great gardens of the colonial south, and in the port cities of the north, where its arrival was an event.

The first advertisement of sweet peas in the colonies, which appeared in a Boston newspaper on March 30, 1760, was an announcement by an enterprising milliner that packets of sweet peas would be exhibited in her shop window along with the latest in spring bonnets and other newly arrived annuals and perennials—a most pleasant custom practiced by the businesswomen of yesteryear.

But the sweet peas of colonial days were still the weedy, reddish-blue or white flowers grown for their fragrance, and not for beauty or show. Sweet peas did not become generally popular in America until the English hybridists transformed them in the late nineteenth century.

Today almost all our sweet-pea seed is produced in the Lompoc Valley in California, where hundreds of acres are devoted to growing commercial seeds for America's gardens. American seedsmen rate the Grandiflora the most popular sweet pea. It bears from five to eight blooms on a stem, comes in a profusion of colors, is resistant to summer heat, and is available in a variety of forms from dwarf bush to climbing varieties.

The modern frilled and ruffled sweet pea bears little resemblance to the fragrant weedy flower discovered by Father Franciscus Cupani, but it continues to bloom with a delicate air, lending a nostalgic charm to summer gardens. It is a hybrid star that for all its latter-day beauty still offers the ephemeral charm of a wildflower and the fragrance of Sicilian fields.

TANSY

In 1846 tansy caused a great to-do at Harvard. Progress had invaded its sanctuaries. "God's Acre," the plot originally set aside for the burial of Cambridge divines and eminent members of the Harvard faculty, was about to be rescued from its pleasant decay; the old coffins of its leading denizens were to be exhumed, and buried again in statelier circumstances. No one knew for sure which grave belonged to what early Harvard president, but it was tansy that identified the first president, who had been buried in July 1668.

The coffin, when it was opened, was filled with tansy that still retained its shape and spicy odor. It was tightly packed about the skeleton, and a circle of it wreathed the skull. The third president, a rival for pre-eminence, had died in December, and when his coffin was opened he was found to be without benefit of tansy, for none, of course, had been available in winter.

The custom of using tansy as a substitute for embalming was carried on well into the nineteenth century by rural New Englanders. It had been the custom in England when the Puritans came to Massachusetts, and they continued to use it here. The practice survives symbolically to this day in many villages, where an armful of tansy is brought by the mourners as a funeral bouquet.

Tansy—a Brief for Health

London medical circles in the sixteenth century were deeply disturbed by a young doctor whose ideas were several hundred years ahead of the times. Dr. Andrew Boorde had taken his degree at Oxford, and spent his next years traveling and studying on the Continent and in the Holy Land. His travels were frequently hampered by a lack of information about roads, inns, and famous places, so he sat down and wrote the first travel guide to Europe.

When he arrived in London to start his medical practice, he was shocked to find many doctors inadequately trained, superficially educated, and given to charging exorbitant fees, which rendered their services unavailable to humbler citizens. He decided that what people needed was a practical guide for home treatment, so he followed his first pioneering volume with a *Breviary of Health*. It was a simple medical herbal, with recipes and directions for the use of plants grown in the dooryard garden or gathered along the roadside. It highly recommended tansy for various ailments, both internal and external, and for its great versatility as a cosmetic, a flavoring, and an insecticide. Undoubtedly his high praise of tansy recommended it to Governor Winthrop when he listed the plants essential for the first herb gardens in Massachusetts. In any event, Winthrop took the *Breviary of Health* to Massachusetts, too, and for two centuries it remained a medical authority in many backwoods households.

Tansy played a versatile role as a home remedy in colonial America. Tansy bitters, made from tansy leaves steeped in a bottle of New England rum, were used as a tonic and sedative. Tansy tea, considered by some the pleasantest of herb beverages and by some a bitter ordeal, was used for ulcers, hysteria, dropsy, rheumatism, constipation, and "woman's weakness," and by wives who wanted to conceive children.

One old book says every woman should bow when passing a tansy plant, not only for its aid in female problems, but also for its cosmetic contribution to a fair face, and for its relief of sunburn and bruises. And in both England and the colonies people wore tansy leaves in their shoes to ward off those dread and common infections ague and malaria.

Tansy's cosmetic properties were relied on by country girls well into the Victorian era. Here is a recipe for beauty from a late nineteenth-century cookbook: "If Maids will take wild tansy and lay it to soak in buttermilk a space of two nights, then wash their faces therewith, it will make them very fare."

Tansy's varied virtues are substantiated by its use in the manufacture of some medicines today. It is listed in the United States Pharmacopoeia as a tea for feverish colds, and for the treatment of jaundice.

Culinary Career

Tansy had a long and varied culinary career. It began in the early days of the Christian church, when the spicy green leaves were added to small Lenten cakes to remind Christians of the bitter herbs eaten by the Israelites at Pentecost. Tansy cakes gradually evolved into tansy pudding, a Lenten custard flavored and colored with the leaves. It must have been more appetizing than it sounds, because the pudding led to omelets and various other dishes flavored with tansy and eaten throughout the year. Many old cookbooks and herbals attest to the satisfaction of these dishes. Gerard says, "In spring time . . . the leaves of tansy newly sprung up with eggs, cakes, or tartes be pleasant in taste and goode for the stomach." Samuel Pepys mentions in his diary for 1666 a supper that included "an omelette flavored with the herb and very good." Izaak Walton provides a recipe, recommended for Lent, involving cowslips, primroses, tansy, and minnows. Some herbalists say that tansy should be included in Lenten cookery to drive away the windiness from eating dried beans, peas, and fish during that season.

In colonial America, and on many farms right through the nineteenth century, the old European custom of packing fresh meat in tansy leaves to prevent decay was observed. Piles of tansy leaves were gathered to rub over the meat and keep the flies away at butchering time. Modern refrigeration negates such age-old lore.

A Short History of Tansy

Tansy is a native of Europe, Asia, and the temperate parts of the Old World. The word "tansy" is derived from the Greek word *athanasia*, immortality, tansy having been the plant the gods employed when they granted eternal life to human beings. The word had been contracted to *tanesie* in old French and German herbals by the thirteenth century.

The Benedictine monks of the abbey at Saint Gall in Switzerland were cultivating tansy in their gardens of medical herbs as early as 1265. By the fifteenth century it was commonly used as a medicine in both France and England, but old herbals caution about its careless use.

It was grown and used in all the English colonies along the Atlantic coast, and soon escaped to the fields and lanes. By

1785 it was listed as a naturalized plant throughout the north-east, and today it is a common weed in the area.

Suspect Tansy, Tansy Benign

Vernon Quinn, a modern herbalist, says in his book *Leaves: Their Place in Life and Legend* that tansy leaves contain a highly poisonous volatile oil, which increases at flowering and seedtime. He stresses, as many old recipes implied, that only young leaves should be used for food and drink. Too strong doses of tansy tea can cause vomiting, convulsions, and even death. But it is grown commercially in some states today, under supervision, for the pharmaceutical business. And after years of neglect as a weed it is coming into its own in florist shops, a growing number of which offer its fragrant sprays and bitter golden buttons to urbanites, who delight in its fresh, intoxicating scent. In November there is a real demand for tansy for dried winter bouquets.

Furthermore, a very special new use has been found for tansy. Like mullein, it is planted in gardens for the blind, so that they may find a certain pleasure and recreation in smelling and handling this age-old herb, whose fragrance is, perhaps, the greatest blessing it bestows on us.

The use of a plant illumines an age.

TEA

Tea was popular in Manhattan years before the English took their first sip. Peter Stuyvesant had introduced it in 1647, when he arrived as the director-general of New Netherland. He came from Curaçao, where he had been governor of the Dutch West Indies, bringing with him a flock of slaves and a chest of China tea: both tea drinking and slavery were already well established in the Caribbean.

He had learned to drink tea at The Hague, where it symbolized the Dutch East India Company's supremacy in Asia. Holland, allied with Portugal, had kept England from the riches of the Far East, including tea, for decades. The Dutch had perceived the value of tea and established their first plantations of it in Java as early as 1606. By 1610 they had imported it into Holland; in 1632 Jacques Bontues, a doctor stationed in Batavia with the Dutch East India Company, published the first description of tea to reach northern Europe. He described it as the universal drink of the Orient; serving tea was not only a traditional social ritual, but—more important to the Dutch—a custom essential to business contracts. Tea became perhaps the key factor in Holland's long success in Asia.

England had little knowledge of tea until 1652, when a chest of it was found aboard a Dutch ship captured off the Channel coast by Admiral Robert Blake. It appears to have been sold to Thomas Garway, a broker of coffee and tobacco; the first advertisement for tea in London appears under his name, and Garway's Coffee House soon offered "a dish of tea." There were, of course, no teapots in England, and tea was served in a bowl and drunk from saucers.

Londoners approached tea with curiosity and caution, and the more puritanical English fired off a salvo of protests against it. Preachers, writers, statesmen, and doctors condemned the new Oriental drink as a threat to health, morals, and public order.

But this situation radically changed when Charles II married Catherine of Braganza in 1662. Her dowry included an alliance with Portugal, with whose support the British East India Company began growing by leaps and bounds. In a gesture more handsome than it may seem, they presented to Queen Catherine a chest of tea from their first China shipment—weighing two pounds, two ounces, and valued at sixty shillings a pound. Catherine had been accustomed to tea drinking at the Portuguese court, and soon made it popular in

London. The English coffeehouses began serving tea, and it aroused such enthusiasm that teahouses were opened in every city and town. Among the most famous of these were London's Vauxhall and Ranelagh Gardens, delightful centers of social activity in contrast to the coffeehouses, which served as clubs for men only.

As teahouses became common, the aristocrats retired to their homes and, led by the duchess of Bedford's example, established the custom of five o'clock tea—a turn of events that proved to be an unexpected spur to the national economy. There was an immediate demand for tea sets in both silver and china, and for teacups, cozies, canisters, chests, strainers, trays, and linens. The ceremonial even led to the invention of the gate-leg table. Tea drinking beside the family fire became so popular that the consumption of gin declined; some public houses were forced to close, and London publicans protested the importation of tea as unfair competition. But tea had come to stay. It was the most profitable item of trade for the East India Company, and tea drinking became an ineradicable English custom.

Tea in America

Introduced so early to New Netherland, tea became the sophisticated social drink there. A choice of China teas—brought by ships of the Dutch East India Company—was brewed in the Oriental manner and served with such other imports as sugar, and with little bowls of saffron and peach leaves for added flavor.

But tea did not arrive in Boston until about the time it arrived in England, and Boston housewives had no idea how to prepare it. They boiled it for an hour, gingerly drank the bitter liquid, and then, with true Yankee thrift, served the leaves as a vegetable seasoned with butter and salt!

In 1664 the British fleet arrived off Manhattan and demanded that the Dutch recognize British rule over the area.

Governor Stuyvesant surrendered, and retired to his farm in the Bowery; New Amsterdam became New York; and tea drinking spread quickly through the English colonies.

The East India Company's profits on tea rose to fabulous heights, but China rebelled at the ever-growing demands for production, and England was forced to look for new sources.

America's Tea Plantations

America appeared to be ideal for growing tea, and on December 12, 1760, the Society for Encouragement of Art and Commerce in the American colonies met in London to lay out plans for establishing tea plantations here. The society did so not without forethought. No less an authority than Linnaeus had grown tea experimentally in Sweden, and had concluded that American plantations would be profitable and convenient.

The society ordered that huge balls of wax embedded with tea seed be sent to each colonial governor, and to outstanding statesmen and gardeners: John Bartram, John Clayton, Benjamin Franklin, Alexander Garden, Cadwallader Colden. But skepticism met England's demand that America grow China tea. After many months of delay, experimental plantations were started in South Carolina, which had the proper climate and a surplus of slave labor. Seed was planted, young trees were set out, and the plantations throve. But before they became commercially profitable the American Revolution broke out, and England was forced to abandon the venture.

Why the Boston Tea Party

British colonial policy, one aspect of which was the imposition of exorbitant taxes without representation in Parliament, became a key factor in precipitating the Revolution. Even after the repeal in 1770 of the Townshend Acts, which had taxed a great variety of imports, the tax on tea remained, and

rankled. Many of the colonists organized to break the monopoly of the British East India Company and to smuggle in tea
from the Dutch East India Company, which readily accommodated them.

When England offered to compromise by sending shiploads of a cheaper grade of tea, including tax, than the tea
clandestinely provided by the Dutch, many colonists considered it an insult to their intelligence and a denial of their
rights. In Boston, members of the Sons of Liberty, led by
Samuel Adams, organized a raiding party, disguised themselves as Indians, boarded the unguarded tea ships in Boston
Harbor on the night of December 16, 1773, and dumped the
tea chests overboard. Similar but less dramatic acts took place
on tea ships anchored in other American harbors. Thus tea
played a leading role in the first act of our War for Independence.

The Revolution was scarcely over when America took
steps to profit from the lucrative tea trade. Elias Derby, a Salem
captain, sailed for China and returned with the first tea imported into the new nation. The venture was an unqualified
success, and over the next decade Derby clipper ships specialized in tea, and imported one-quarter of all that was sold in
America. The profits of a single voyage ran as high as
$100,000, and Derby retired as America's first millionaire.

In the meantime the British East India Company, thwarted
in its plans for American tea plantations, sent the plant explorer Robert Fortune to collect tea plants in China and establish plantations in India, which England now controlled. This
venture proved so eminently successful that the young United
States government invited Fortune to come here and supervise
the re-establishment of tea plantations in South Carolina. Fortune arrived in 1858, tea seeds were sown and trees transplanted, and success seemed assured. But three years later the
Civil War began, and by the time it was over the south was
impoverished and slave labor was no longer available; so the
plantations were abandoned.

Early in the twentieth century experimental tea plantations were again attempted in various parts of the south and
southwest, by the Department of Agriculture and by private
investors. Labor-saving machinery was devised, but it is
mandatory that tea leaves be picked by hand, since those
nearest the top of each plant provide tea of the highest grades;
labor costs therefore remained a paramount problem. An
article appearing in *Meehan's Monthly Magazine* in 1901 implied that the problem had been solved by training child labor
to pick the tea, and an edition of the *New International Encyclopedia*, published six years later, reported that the out-
look for large tea plantations in the southern United States
was good. But new labor laws soon canceled these possibilities, and commercial tea plantations were finally abandoned in
the United States.

Tea History—Fact and Fiction

An ancient myth has it that in the year 2737 B.C., the Emperor
Shen Ming was boiling water on a brazier in his garden when
some leaves from a tea plant blew into the pot, and soon gave
off a pleasant aroma. The emperor tasted the liquid, found it
delightful, and recommended it to his court. After this royal
introduction tea drinking spread throughout Asia, and down
the centuries a great number of customs and ceremonials
gathered about its brewing and serving.

Tea, *Thea sinensis*, is an evergreen related to the
camellia. Most authorities believe it is indigenous to the state
of Assam, in northeast India, and possibly to parts of China
and Japan. The cream-colored blossoms are fragrant, the
leaves are dark green; and cultivated trees are kept to a height
between three and five feet by pruning to facilitate picking,
although in their native state they may attain thirty feet.

All tea leaves are dried and roasted after picking. The
green teas of China and Japan are produced by firing the
leaves immediately. The black teas characteristic of India and
Ceylon result from a fermentation of the leaves before firing;

leaves for the oolong tea of Formosa (Taiwan) are fermented only partially.

Tea is an international drink, and in some parts of the Orient is still used as a kind of vegetable relish, as it probably was at the very beginning of its history; some Africans chew the raw leaves for their intoxicating properties. Buddhists from China introduced the tea ceremony into Japan, as a social custom with religious overtones, in the fifteenth century.

Until fairly recently, many Britons drank tea immediately after waking up in the morning, and again at breakfast, and as part of "elevenses" at midmorning. Its finest hour probably still arrives in late afternoon, when it is sometimes served almost as ceremoniously as in modern Japan. Now there are signs that it may be yielding, at least a little, to a renewed fondness for coffee, but the United Kingdom still consumes one-third of the world's annual production of tea.

Generally speaking, we approach tea more casually in the United States, and have welcomed the convenience of tea bags and even of instant tea, including iced-tea mixes with built-in sugar and lemon. This may help to account for the fact that we stand near the top as importers of tea, along with Australia, Canada, Russia, and the Netherlands.

More tea is consumed by humankind than any other drink but water, because it is the least costly and the most readily available stimulant on the planet—and perhaps the most agreeable.

THISTLE

When the Vikings sailed down from the Northland to raid Scotland's wind-swept shores—so say old tales—the Scots piled the beach head with thistles, and the laird gathered his clan into his smoky stone keep, where they waited. Night came. Suddenly curses and screams of pain tore the silence as the Norsemen, barefoot in sandals, leaped onto the thistle-strewn beach. Warned of their approach, the Scots drove them back to their ships, and thus the thistle became the emblem of Scotland and the source of her motto, "Touch Me Who Dares."

History

The thistle is a Eurasian weed that has been associated with man from the earliest days, that has gained many folk names and been put to many uses. There is much confusion about the classification of the endless varieties, but thistles may be divided into three general groups: the *Carduus* or holy thistles, the *Sonchus* or edible thistles, and the *Centaurus* or garden thistles.

The *Carduus* thistles, native to the Mediterranean, were used as medicinal herbs long before the Christian era, but the religious enthusiasm of the Middle Ages ascribed their beneficence to the saints. The white spots on the leaves of the milk thistle, *Carduus marianus*, were said to have been caused by Mary's milk dropping upon them. Centuries of miracles are ascribed to the milk thistle; Charlemagne used it as a cure for the plague, and Shakespeare refers to its medical virtues in *Much Ado About Nothing*.

The Benedictine thistle, *C. benedictus*, named for Saint Benedict, who founded the order that bears his name, was grown in early monastery gardens and credited with miraculous cures for smallpox. The Venerable Bede introduced it to England in the eighth century, and early herbals recommended it for healing sword wounds. Even in the eighteenth century, John Hill, a distinguished botanist who was also an early promoter of natural foods, recommended thistle bitters as an excellent "stomachic."

Though the sacred thistles were used primarily for medicines, they were also considered edible, and their roots were used as an equivalent for salsify until modern times. But it is the *Sonchus*, or sow thistle, that has been a dependable food for untold ages in Africa, Asia, and Europe. Both its fleshy taproot and its broadly rounded leaves were used as pot herbs and in salads. A thirteenth-century herbal highly recommends a diet of sow thistles "to prolong the virility of gentlemen." Gerard, who reports that the poor stuffed their ticks and pillows with its abundant down, also cautioned that deceptive upholsterers grew rich by diluting costly goose or duck down with thistle.

The *Centaurus* thistles take their name from Chiron, the centaur who instructed Achilles in the use of herbs. (See Yarrow.) They include the knapweeds, the star thistles, and the cornflowers, which were among the thistles grown in every English garden as home remedies for various ailments, and which were brought to New England by the Puritans.

The Thistle in the United States

Few records of the introduction of thistles to the colonies exist, but John Josselyn says that several varieties grew in Puritan gardens in 1637. He relates that when he visited the Indians and found many of them suffering from "stuffing of the lungs," he made them a tonic from leaves of the sow thistle, catnip, anise, and fennel—all European herbs he had gathered in the settlers' gardens. Here the bluebottle or common cornflower, *Centaurea cyanus*, was also cultivated, as a home remedy for inflammation of the eyes, and for jaundice.

The garden thistles soon escaped to the cleared fields and meadows of the colonies. Others were unintentionally introduced in the fodder brought for the first cattle and horses. Among these secret invaders the bull thistle has become an intense problem. Even the striking beauty of its large purple heads cannot compensate for its annual ruination of thousands of acres of America's cultivated and pasture land. The Canadian thistle, another deceptive beauty and perhaps the best-known member of the tribe, is not in fact native to Canada, but was naturalized from Europe. It has spread across the border and is now a noxious weed overrunning the whole northern half of the United States. The thistle invasion has in fact been going on for three centuries, in many guises. There are the familiar plumed thistles, pasture thistles, swamp thistles, and lance thistles among the varieties—almost 120—that have taken up land in the United States. The distinguished American naturalist Thomas Nuttall says there is only one native thistle, *Cirsium americana*, and even this was probably brought to our southwest by early Spanish settlers.

Botanical Bits

The thistles are characterized by spiny needles that cover the stems, the alternate leaves, and the terminal discoid flower heads set in cups of spiny bracts. These heads vary in color from creamy white to royal purple; they measure from about one to two inches across, and consist of compact, tubular, perfect flowers that give them a fluffy, pincushiony look.

Thistles ensure their possession of open land whether they are perennials, biennials, or annuals. The perennial Canadian thistle, with its creeping horizontal roots, and the biennial bull thistle, with its long fleshy taproot, are also blessed with plumed seeds, which make their eradication almost impossible. The *Centaurus* thistles appear as enduring annuals, and possess bristly seeds that enable them to thrive in such diverse areas as abandoned lots in city slums, highways, and Florida swamps.

The Uses of Thistles

But the thistles also render service to our modern world. They are the favorite food of the goldfinch, and so many other birds enjoy them that they have become a popular commercial bird food. They also provide food for wild bees,

and protection for some butterflies, which lay their eggs in
the axis of the bristly leaves because marauding ants cannot
navigate the stems and cattle will not eat them.

The large-flowered, Old World globe thistle, *Echinops*, is
a handsome perennial grown by many American gardeners.
But it is the Scottish thistle, *Onopordum acanthium*, that still
grows at Holyroodhouse, and those who have seen it there
will never doubt that it was especially designed for gardens
everywhere.

TIGER LILY

What subtle Chinese wisdom appointed the tiger lily the inscrutable potentate of garden flowers? For this it surely is, as
ancient lore assures us.

A Chinese herbal entitled *A Record of Precious Things*
states simply that the tiger lily is the prince of plants, dedicated to the comfort of man. A Korean legend asserts that it
really is a tiger under an enchantment—one who protected a
hermit priest, ages ago, and in return was given eternal life on
the condition that he continue to serve not just one man, but
all humankind. Old Japanese garden manuals suggest a more
sinister magic, listing it as "Onigura," the ogre lily.

I first met the tiger lily in my grandmother's garden,
growing under the sour cherry tree. No one seemed to remember when it had been planted, but there it stood—alone,
defiant, and provocative, made the more mysterious by the
tiny brown bulblets in the axes of its leaves.

After many summers of hesitant observation, I was
somewhat reassured by *Through the Looking Glass*, in which
I found that Alice, too, had met a tiger lily in a garden and
found it condescending and unapproachable. When I began
actually studying it, I became convinced that there is more
than a little truth in all the ancient lore.

Historical Background

The tiger lily is a relatively recent arrival in the pleasure garden, although it has been cultivated as a food for 2,000 years. In China, Japan, and Korea, its bulb is as basic today as the potato is for us. The highly nutritious scales on the bulb are peeled, cooked, seasoned, and eaten by millions, and also taken as a spring tonic. These facts would be incidental if lilies in general were cultivated as a food crop. But they are not: they have stubbornly resisted cultivation, and of the hundred or more species native to the Northern Hemisphere, relatively few have submitted to man, and these principally as ornamentals. Among the exceptions are day lilies (which see), the star and musk lilies of China, and the Madonna lily that served Europe as a medicinal herb throughout the Middle Ages. Though all wild lilies are collected for food in the Orient, the tiger lily is the only one that has served as a dependable food crop for so long.

Another of its remarkable characteristics is its integrity. Although it has been exploited as a food through the ages, it has remained essentially unchanged, which is the more amazing when we reflect that virtually every vegetable in today's markets has been "improved" in one way or another—made larger, given a more appealing color, conditioned for shipping and storage. The tomato, potato, and carrot, for example, are radically different from their counterparts of a century ago.

Oriental Lilies in Europe

The first record of the tiger lily in Europe comes from Engelbert Kaempfer in 1684. He was a young German surgeon with the Dutch East India Company who spent ten years collecting new and rare plants in the Orient. (See Hydrangea.) One of the first plants he sent from Japan was the tiger lily, and later, in his book *Amoenitates Exoticae,* he included a colored sketch of the tiger lily.

Unfortunately the plant itself was lost to cultivation in Europe after the fanfare that greeted its introduction, and little was heard of it for another century. But Kaempfer's book had a lasting impact because it described many hitherto unknown Oriental plants, including magnolias, clematis, azaleas, and hydrangeas, and aroused a widespread interest in importing them.

In 1804, a Captain Kirkpatrick of the British East India Company returned from Canton with a shipment of plants collected by William Kerr, a plant hunter who had been sent to China by Kew Gardens. The collection included some tiger-lily bulbs that were successfully propagated at Kew, and *Lilium tigrinum* became one of the first Chinese lilies to be permanently established in Europe. So it moved from Asia's kitchen gardens into Europe's flower gardens, the precursor of a flood of Oriental lilies that poured in from the ardent plant hunters of the nineteenth century—Carl P. Thunberg, Robert Fortune, Augustine Henry, George Forrest among them. By 1867 the tiger and many other choice Oriental lilies were established in the botanical gardens of Paris and Brussels.

The Tiger Lily in America

The tiger lily made a dramatic entrance into America, too. It arrived on one of the first clipper ships from China, a coincidence that has particular significance because it symbolized America's independent trade with the Far East. After the launching of our swift China clippers in the 1830's, nurserymen could import floral wealth directly from the far places of the earth and propagate it here instead of relying on European garden and orchard stock.

The tiger lily was soon growing in Salem and Boston, and in many port cities along the Atlantic coast, including Baltimore and Philadelphia and Charleston, and in the plantations of tidewater Virginia.

After the Civil War it began its long trek west with the

pioneers, obligingly scattering its tiny bulblets wherever they built their cabins. This form of self-propagation has proved so efficient that the tiger lily now grows wild across the nation. Along the northern reaches of the Pacific coast and up into the Cascade Range, it appears in special and spectacular abundance.

Why the Tiger Lily Is Unique

The tiger lily continues to defy botanical science. It will not change its spots.

Lily breeding is having a great revival under the supervision of today's hybridizers, but the tiger lily has resisted their efforts to break down its identity, and it refuses to bear seeds. It will feed man, luxuriate in his garden, and beautify his wilderness, but it remains aloof—an Oriental mystery.

Staff plant breeders of the New York Botanical Garden set out to investigate this tenacity. They discovered that the tiger lily is a natural triploid and has three sets of chromosomes in the nucleus of its cells. This was surprising, because lilies are normally diploids, with only two sets of chromosomes.

Instead of developing seeds in the ovaries of its flowers, the tiger lily bears tiny bulbils in the axis of its leaves. These fall to the ground, put down a slim sustaining root, and send up a strong new plant. This is an efficient method of reproduction, but it does not result in accidental hybrids or mutations. Modern hybridists were challenged by this unique simplicity.

Experimenters treated the tiger-lily seeds and seedlings with colchicine solutions and produced tetraploid plants. But they proved sterile. Others hand-pollinated the tiger lily with pollen from diploid lilies and proved they could bear seeds, but the results produced only a colorless, deficient hybrid lily so unlike the brilliant tiger that experiments to discipline the individualism of this superb plant were discontinued.

The ancient enchantment still holds, and I, for one, am glad. The tiger lily still nods his princely head above my tall white phlox—inscrutable, uncompromising, unchanged after 2,000 years.

TREE OF HEAVEN

Spring and summer gradually transformed the parking lot. The abandoned building on one side acquired a hedge of seedling ailanthus trees all along its foundation, and the broken wire fence a screen of bindweed vines bearing thousands of white blossoms. Milkweed, growing lushly beside the fence, entirely covered the broken bottles and rusted cans, and when it bloomed, its purple flowers were topped with fluttering orange-and-black monarch butterflies. The old tree of heaven put out long fronds of leaves, and by midsummer its extravagant clusters of crimson-and-green-gold seed pods waved gently in the wind, lending an air of Oriental mystery to the spot.

How did this dramatic Chinese tree arrive in our cities, and why do contemporary city planners denigrate its beauty and utility?

The first tree of heaven to arrive in America was brought from Paris in 1784 by a Philadelphian named William Hamilton, who planted it on his famed estate, Woodlands, outside the city. He had seen the tree growing in the garden of the royal physician to Louis XVI, and learned that it was a great rarity, grown from seed sent to France in 1751 by the missionary Pierre d'Incarville, who was the first European to investigate the natural world in northern China.

As if some higher wisdom had ordained the tree's future service to the West, its seed had escaped the theft, shipwreck, or neglect that had seemed like a curse on d'Incarville's collections. This encouraged him to send seeds to the Royal Society in London, and to Phillip Miller at Chelsea Gardens. Today a great tree of heaven, a descendant of those seeds, grows in the botanic garden of Cambridge University.

Hamilton's coveted seedling from the Paris tree grew rapidly on his Pennsylvania estate, and was admired by many Americans, among them William Prince, of Long Island. By 1820 Prince was importing trees of heaven for sale, and recommending them widely for landscaping American estates and parks.

It was his enthusiasm that persuaded the growing metropolis of New York to use it as a street tree, and it began to replace the native maples, lindens, and oaks. Noting its beauty, vitality, and rapid growth, Philadelphia, Baltimore, and the District of Columbia soon made efficient use of it, too.

In rural areas the discovery that it was exceptional for firewood, fencing, and farm equipment encouraged its wide cultivation. And about 150 years ago Americans learned that the leaves could provide sustenance for one of the silkworms, *Attacus cynthia,* from whose cocoons a serviceable pongee silk is made. Many southern planters set out acres of the tree, silkworms were imported and thrived on the foliage, and dreams of wealth were nearly realized. But the cost of labor and the lack of skilled workers made the manufacture of silk unprofitable. So the trees of heaven, like the mulberry trees that had also been imported for the production of silk, were considered of no further commercial value.

By mid-century, the tree of heaven had spread of its own volition from Massachusetts to the Mississippi, and from the Deep South to Texas. But its very vigor undermined its popularity. Cities found the maturing trees difficult to control. Their insistent spreading roots sent volunteer trees up through sidewalks and the foundations of buildings, and along roadsides where nobody wanted them. When they began to

bloom, millions of inchworms dropped from them onto
pedestrians, and many people were allergic to their pollen.
Worst of all, the male flowers gave off an offensive odor, as
even d'Incarville had observed—his own name for the tree,
indeed, was "stinking ash." The inchworms were extermi-
nated, the odor was circumvented by planting only female
trees, and analysis proved that the pollen was far less irritating
than that of the linden and the poplar. But all efforts to improve its image were futile, and the tree of heaven was
banished from America's metropolitan areas.

Nevertheless, it was here to stay. Although the New
York City Parks Department uproots thousands of volunteer
trees of heaven each year and no permit for planting them has
been issued for decades, it is estimated that more than a half
million grow in the city alone, and thousands more line the
highways across Long Island. And what is now the oldest tree
of heaven in America is still the showpiece of a Long Island
estate. Planted in the last century, when it was considered an
Oriental ornamental, it stands over eighty feet high, and has a
trunk over seventy inches in diameter.

Mostly the tree of heaven has retreated to tenement areas
and abandoned lots, careless of the impoverished soil. Where
no other green thing can thrive, its great fronds of leaves
provide shade and beauty. Meanwhile the conventional street
trees in our cities are subjected to a form of torture. Their
amputated roots are disciplined to concrete, their leaves
struggle for sun in the canyons created by high-rise buildings,
and soot, sulphur, and noxious gases interfere with the vital
process of transpiration. They turn yellow, sicken, and die.

Ailanthus in City Planning

The tree of heaven is almost impervious to pollution. Its great
compound leaves, composed of eleven to forty-one leaflets,
achieve maximum exposure to sun and wind and retain far less
soot than simple leaves do. Its aggressive spreading roots

absorb a maximum amount of water, helping it to remain healthy. In Europe it has been widely used in replanting bombed-out cities, and has grown with amazing vigor where no other tree would grow at all. It is also much grown in Paris and London, and in cities in Italy, Germany, and Spain. Many Europeans in rural areas utilize it as an effective boundary hedge. They plant its seeds in rows, and the saplings—which grow from ten to fourteen feet in a summer—are cut back to a desired height each spring. The result is a beautiful and durable fence.

In Peking, the northern limit for what is really a tropical tree, and in many other Far Eastern cities, it is planted along many streets for its beneficent shade. The Chinese call it "ch'u," but it is native to all the warm regions of Asia and the South Seas, including the Moluccas, where it is known by its Amboina name, *aylanto*. This word, which implied a tree that reached to heaven, was corrupted by European naturalists, who gave the tree the official botanical name *Ailanthus altissima*.

While Europe and Asia utilize the endurance and adaptability of this ancient tree to the modern urban problems of pollution and sunless concrete streets, it remains an outcast in America. But isn't it a perfect tree to play a part in our urban renewal?

Early History and Use

Fossil fruits and leaves found in tertiary beds in Montana are so nearly similar to those of the modern tree that there is no doubt that *Ailanthus altissima* grew there before the ice age destroyed it 60 million years ago.

There are only seven species of this tree of the Simarubaceae family, and the only near relative found in the West today is the paradise tree, *Quassia amara*, of the Caribbean. It yields quassin, used in the treatment of malaria; the Carib Indians believed that water drunk from a cup made of quassia wood could cure chills and fever. The tree was named for a

Surinam black man, Graman Quassi, who discovered its medicinal virtue in 1730 and extracted from its bark a drug that is a remedy for pinworm, a dread parasite of the tropics.

In its youth the ailanthus suggests our native sumac, but it soon matures into a dominant tree. Panicles of yellow and green flowers develop by midsummer into handsome clusters of green-gold-and-crimson samaras, or long twisted seed pods. Each pod contains a single black seed, which is carried great distances by the wind, for it is very light: there are 17,000 of them to a pound. It is as dramatic in winter as in summer. Its naked limbs make a strong, emphatic, abstract pattern, and they are covered with huge shield-shaped leaf scars, marked by nine large bundle traces and topped by plump pubescent leaf buds. Thus it awaits spring, when it will once again provide beauty and a refreshing flickering shade as it prepares the seeds that will ensure its perpetuity.

What other tree can compete with this Asian immigrant in our brave new world?

TULIP

The story of the tulip's introduction to Europe, and of the tulipomania that subsequently swept the Continent, has been told again and again. But a mystery still clings to these shapely, shining flowers. We Westerners know little about them in their native state, although they are indigenous to the North Temperate Zone from the Mediterranean eastward to Japan, grow wild in the Crimea in great profusion, and must have been cultivated by the Persians centuries ago. But they are nowhere mentioned in Greek and Latin herbals, and as far as Europeans are concerned seem to have sprung into being— full-blown, sophisticated, and mature—in the sixteenth century, in the garden of the Turkish sultan. Ogier Busbecq beheld them there around 1550, and "at a great price" secured seeds and bulbs and sent them to Vienna, to the Flemish botanist Charles de Lécluse, who then had charge of the Imperial Botanical Garden.

Five years later tulips were growing in the gardens of the great Fugger family's palace in Augsburg. These powerful merchant princes greatly admired tulips, which began to appear—doubtless through the Fuggers' influence—in gardens of the wealthy in Holland and Belgium, France and England, and even in Russia. But it was not until Lécluse was appointed professor of botany at Leyden University in 1593 that tulips attracted general interest. He took a store of seeds and bulbs with him, and the prohibitive price he asked for them made them the more intensely coveted. The almost inevitable result, so the story goes, was that Hollanders stole both seeds and bulbs, from which the first tulip gardens sprang up in the Netherlands. By the early years of the seventeenth century they had spread throughout the nation, and the bulbs were being sold at a great price to an increasingly avid public.

As long as tulips in their original, solid-colored form remained a relative rarity, they were traded in the financial market much like any other commodity of the era. As they became more common their price fell, but there was a mounting obsession with tulips that were sports or freaks, exhibiting a striking if not necessarily beautiful change in pattern that may have been the effect of a virus rather than a true mutation. Not just the bulbs themselves, but formulas for forcing the solid-colored tulips to "break" and produce striped or feathered or marbled forms, were sold for large sums.

Aristocrats, farmers, merchants, sailors, workmen, and servants were caught up by tulipomania, trading their land, livestock, crops, even food and clothing, for tulip bulbs considered priceless or capable of producing priceless descendants. A single bulb of the multicolored sport named "Semper Augustus" brought 5,000 florins plus a carriage and a pair of fine horses; and other single bulbs, the equivalent of several thousand of today's American dollars.

The gambling in tulip futures became so intense that many farmers grew them to the total exclusion of food crops; actual suffering ensued, and the impoverishment of many humbler Hollanders. For a time tulipomania also infected the commodity markets of London, Paris, and Vienna, though never to such an extent as in Holland. There the government at last stepped in, and set a strict limit on the price of any single bulb. The craze, by then, had almost run its course; supply more and more outran a waning demand, and by 1637 the bottom had fallen out of the market.

The collapse of tulipomania brought ruin to many small farmers and speculators, and an economic depression throughout the Low Countries. But as time passed and the economy slowly recovered, Dutch farmers discovered that there were many people of moderate means who, in a perfectly reasonable way, yearned for beds of bright tulips in their spring gardens. Down the years the raising and exporting of bulbs became a solid national industry, with a steadily expanding market, which the Dutch in Holland continued to dominate, even after commercial bulb growing spread to other nations, notably our own, more often than not under the aegis of Dutch immigrants. By the late 1930's Holland was exporting 100 million tulip bulbs to the United States annually.

A sad, bizarre sequel to the industry's high success occurred in the World War II years, when starving Dutchmen were forced to eat their stock of bulbs. This, though, was an economic rather than a gastronomic disaster: numerous experimental samplings have established that tulip bulbs are quite as edible as their cousins the onions, and there are those who maintain that they are in fact delicious.

Tulips in America

New Netherland was being settled at the very time tulipomania was rampant in Holland, so it isn't surprising that tulips grew in the first gardens in Manhattan.

Adriaen Van der Donck came to New Netherland in 1641, with the Dutch title jonkheer; the city that now occupies the great tract of land granted to him is called "Yonkers" for that reason. It is from his detailed account of horticulture in the Dutch colony that we know tulips were among its garden flowers. He was the man who persuaded Peter Stuyvesant, when he arrived to be governor, to lay out the garden at the Battery that became Stuyvesant's greatest pride. So tulips grew there, part of the largest collection of foreign flowers in the early colonies.

Tulips were grown in Williamsburg, too, and at Mount Vernon and Pennsbury and Saint Mary's, brought by English ships from Europe's nurseries or by individual colonists—for nothing was easier to transport than flower bulbs. Their popularity among the Pennsylvania Dutch is suggested by the tulip design that appears so frequently in the products from their potteries that it is often called tulip ware. And the descendants of the Hollanders who emigrated in such numbers to western Michigan a century or so ago still hold a tulip festival every year in the town called Holland, a tulip-growing center.

The tulips brought to the colonies were the old solid-colored and striped ones, perhaps along with some parrot tulips, which had appeared in Europe about 1620, and some white lady tulips, which were popular in England in the first half of the seventeenth century. But there were no Darwin tulips, because they were not developed until the late nineteenth century.

Today the tulips available constitute a literally dazzling array of types and varieties. The hybridizers have been hard at work in recent decades, and have wrought dramatic changes. The range of color has been greatly extended, and also rendered more subtle. Among the numerous varieties, the flowering season extends for about two months—in the northeast, from early April through May. The stems of several strains have been greatly strengthened; the wide variation in shape includes both ruffled and branching types; there are even, at last, tulips whose blossoms are truly fragrant. Most gratifying of all, perhaps, is that some of the new tulips, given proper care, come close to being truly perennial instead of blooming satisfactorily for only a season or two.

History Extras

The Turks called tulips "lale" and used them as love tokens, and they became entwined in the myths of that region. Our name for them derives from the Persian word for turban, *dulband*, which became the Turkish *tuliband*, Latinized in turn to *Tulipa* for the plant's botanical name. Traditionally the tulip had little or no perfume, and it has never served as a medicinal plant, or for dyeing or any other domestic purpose. Throughout its history, simply to bloom and add its particular splendor to spring gardens has sufficiently justified its existence. John Parkinson, in his botanical work *Paradisi in Sole Paradisus Terrestris*, published in 1629, sums up the reasons for the tulip's centuries of popularity: "There is no Lady nor Gentleman of any worth that is not caught with their delight."

TUMBLEWEED

The great sparkling balls of tumbleweed rolled across the prairie, beautiful and mysterious. The terrified child in the lone cabin watched her exhausted parents beating out the sparks wherever the wind dropped the smoldering globes on the dry grass growing nearby, or on the parched field of grain. The wayward wind had carried the tumbleweed for miles, dropping its dried skeleton of leaves and branches near a roaring prairie fire, picking it up again before it flamed, and dropping it again, to touch off another grass fire. It became the witch grass, the destroying angel of the pioneer.

Even when my grandmother was ninety her terrifying tales about tumbleweed, which had destroyed her father's crop and left the family homeless on the prairie, still made it a vivid and malign presence in our New England childhood.

A Russian Immigrant on America's Plains

The tumbleweed, *Salsola kali tenuifolia*, arrived in the United States in the early 1870's with a group of Russian immigrants who settled in South Dakota. Their stated purpose was to raise flax on the plains; tumbleweed or Russian thistle came along as a stowaway in their hand-threshed flax seed. The cultivation of flax had only a limited, local success, but tumbleweed soon won international fame as the pest of the prairies, and played a key role in the destiny of many settlers.

It was first noticed in 1873 in Bon Homme County, the center of the Russian settlement; twenty years later it had spread south through sixteen states and north through thirteen Canadian provinces. Its dominance reached a climax in 1930, when the great drought killed many crops, and many less aggressive weeds as well, enabling tumbleweed to extend its territory throughout the Dust Bowl, including much of western Texas. Since then, men and nature have worked to control this immigrant plant.

Biography of a Tumbleweed

The tumbleweed has collected many common names, among them "wind witch," "Russian cactus," and "saltwort"—which last it takes from its preference for salty alkaline soil.

The typical tumbleweed is a great globe with wiry branches, 12,000 small spiny leaves, and minute greenish-pink flowers. It demands from three to five feet of space, and its shallow root systems may spread even more widely. Marvelously adapted for invading the American prairies, it arrived when pioneering was at its height, competed with the settlers, and often won possession of the land.

The virgin prairies, whose endless grasslands flowered with scarlet Indian paintbrush, golden bitterweed daisies, wild onions, coreopsis, and coneflower, made the pioneer's eyes dance. Coming from the east, where forests and apparently self-propagating stones had to be dealt with before he could farm, he translated the prospect into those amber waves of grain that we sing about in "America the Beautiful." With backbreaking labor he plowed the deep layers of prairie grass, and saw his young garden flourish, just as he had imagined. But then, out of nowhere, the tumbleweed came, carried on the wind, and, blown about for a month or more, sowed its tiny black glossy seeds—20,000 of them to a plant—over a space ten miles square.

It flourished in the hot summer sun, achieving an eventual weight of seven pounds. It remained green while the pioneer's flax, alfalfa, and grain wilted and died, because it needed less than half the water of most food plants. Its wide branches crowded out what grain remained, and when autumn came, with its roaring prairie winds, the tumbleweed broke off from its short brittle stem and went bowling and bounding away, scattering its seeds in installments. Those near the broken stem were loosely held in place, and fell nearby. As the wind picked it up and dropped it, over and over again, other seeds ripened and fell, until its ragged remains came to rest in a creek bottom, or perhaps piled up against a barbed-wire fence, where it was eventually covered by a dust storm, from whose residue its remaining seeds would sprout in the spring.

The traveling dried tumbleweed was most feared by the pioneer in its role of natural arsonist, carrying in its sere skeleton sparks from a prairie fire miles away. It was also a prime hazard to the early railroads: clogging creeks, piling up on cuts and railroad embankments, it collected the blowing dust, covered the tracks, and caused delays and accidents.

But the railroads profited by nature's lessons. When dust storms increased as settlers plowed up more and more of the prairie grass that had held the soil in place for centuries, barbed wire strung at strategic spots caught the tumbleweed, which piled up against it and in turn caught the blowing dust and formed a barrier that kept the tracks relatively clear.

A Reverse Pioneer

The immigrant tumbleweed has become a pioneer plant in the restoration of America's grasslands. Though it was so often a destructive agent, it was also among the first plants to appear on fields rendered bare and barren by the plow and harrow and the prairie winds. In the shade of its wide branches insects, birds, and small animals found shelter. Its infinite seeds fed returning quail, gophers, and squirrels, and its young plants fed deer and antelope. Its spreading roots held the soil, enabling wide-leafed annuals and perennials to grow, and to retain the moisture in the ground, thereby reducing dust storms.

Unrestricted, tumbleweed sows itself so thickly over the plains that it looks like a green carpet in the spring, and its autumn red is one of the spectacles of American desert country. It is from this habit that it takes its botanical name *Salsola*, the Persian word for "carpet," and *kali*, from the Arabian *al-qaliy*, meaning the ashes of saltwort.

But it is because of—rather than despite—its prolific seeding that it does not overwhelm the grasslands. It is disciplined by its own fecundity; the young plants, which often sprout a thousand or more to a square yard, compete so intensely for space that only the strongest live to maturity.

There are several other plants called "tumbleweeds," including the tumbling mustards or hedge mustards, and the tumbling pigweed, *Amaranthus albus*, which rivals the Russian thistle, bearing on one plant enough seeds—which are viable for twenty to thirty years—to sow an acre of ground. Also referred to as tumbleweeds are the native buffalo burr and the cyclone tumbleweed, which is confined to the far west.

The invasion of the immigrant Russian tumbleweed had a vital effect on the settlement of the west. Although it destroyed the hopes of many pioneers, it has become a constructive agent in the healing of the grasslands. So it is another of those symbols of the interdependence of all living things. Man cannot control nature, which disciplines animals and plants alike to create an environment for the renewal of life and abundance.

UGLI

American fruit stands in our northern cities are taking on a new look. Oranges, grapefruit, and bananas are being joined by several new fruits from the Caribbean, among them the plantains, and the ugli, a native of Jamaica.

When you first see an ugli you decide it's aptly named. Then you read the price tag—fifty to seventy-five cents each—and your curiosity is stirred. What intrinsic pleasure does it offer to overcome its appearance, and a price that would be high for any citrus fruit? Even in Florida I found it tagged at forty cents.

But this wizard from the Caribbean has cast its spell—and you buy one. It looks like an overripe degenerate grapefruit gone soft. Its rind is thick and tough, and it is full of seeds. But its taste is interesting and uncommonly refreshing. You are beguiled into appreciation.

The history of the ugli is clothed in mystery, too. It appeared about fifty years ago in a citrus grove in Jamaica, apparently a cross between a mandarin orange and a grapefruit. This would place it in the tangelo family, but its actual parentage has never been proved, and its propagation has been entirely by budding.

It was exported to England in 1934, after it had become popular with native Jamaicans and English tourists. It proved to have excellent shipping qualities, and the English rated its eating qualities excellent, too. So, despite its unprepossessing appearance, it soon won a place as a dessert fruit in the English markets.

Then it invaded Canadian fruit stands, and again found instant success, perhaps because many Canadians winter in Jamaica and grew familiar with it there. It was generally called the "ugly," and the name seemed so appropriate that it was copyrighted by the Jamaican exporter G. G. R. Sharp, who had cashed in on its superior eating and shipping qualities.

The price the ugli commands and its growing popularity must surely arouse the interest of any citrus grower in our south. But we now have no ugli trees in the United States, and because of the restrictions on citrus imports we are not likely to have any. United States quarantine regulations prohibit entry of budwood of any citrus variety, in order to protect our extensive groves against disease and insect pests.

Of course seeds of ugli may be brought in, and many growers have tried starting trees from seed in the past. All trees so far grown from seed have refused to bear ugli fruit. Shortly after the first crops were proving commercially successful, and before the United States restricted its cultivation, Florida growers did import budded wood, and experimented in a limited way with ugli groves. But only a few trees ever produced, and these soon appeared to be ailing from some citrus virus. Moreover, the crop was small and the fruit so inferior in quality that it offered the Jamaican ugli no competition.

Citrus varieties often exhibit a disinclination to adapt happily to a new environment and a different soil, and these may have been prime factors in dooming Florida uglis. Whatever the cause, they fell victim as well to United States quarantine laws, and to frosts that have killed off the last of the experimental trees. In 1977 uglis were still appearing on the fruit stands of grocery chains, competing in price—if not in beauty—with the best large grapefruit.

It seems likely, then, that Jamaican growers will enjoy an ugli-fruit monopoly for the foreseeable future. According to the *Citrus Industry*, their production on that island is increasing.

VERBENA

Today's garden verbena is a truly modern flower, unknown a century ago. It typifies our nation as no other flower does, because it is the product of diverse immigrants. Like our people, this happy hybrid has retained the sturdiness and adaptability of its native land, but has thrived in the generous atmosphere of North America, and grown more beautiful and varied.

The first verbenas were discovered four centuries ago when the Spanish conquistadors tramped over the Andes, and the irresistible fragrance of lemon verbena captured their curiosity. The Spanish padres who traveled with the conquistadors and who were trained botanists, insisted that it was a relative of the European vervains. They learned that the Indians used it to make a fragrant tea, and they gathered quantities of its seed to send home for cultivation in monastery gardens. As a result, verbena tea enjoyed some popularity in Spain during the sixteenth and seventeenth centuries.

But other verbenas, too, grew in the New World. A purple variety was discovered in Peru, and white-flowering verbenas proved to be indigenous to all sections of South America. The Spanish in Buenos Aires and the Portuguese in Brazil both claimed credit for the scarlet verbena, which demonstrated again that nature has no interest in man's political boundaries.

Despite the abundance and variety of this lovely new flower, Europe heard little of the verbenas, or of many other newly discovered plants, for more than two centuries, because of the Spanish policy of interdiction.

But when the eighteenth century opened with an intense interest in New World plants, various forces combined to promote the verbena's fame. In 1785 one of the first New England clippers, homeward bound from the Orient, stopped in Chile before sailing round the Horn. The captain, who was an ardent gardener, discovered lemon verbena there, and took home to his Salem garden what is said to be the first verbena of record to reach the United States. Its fragrance and beauty were much admired, and before long many other American ships were collecting verbenas from South American ports.

Verbenas wound their way into the hearts of colonial gardeners because they provided both color for the eye and satisfaction for the nose, as well as a savory herb tea that relieved colds and fevers in the pleasantest way imaginable.

During the next century, verbenas spread throughout the eastern states, as prolific and as varied as they had been in South America.

Later Plant Hunters
In 1825 James Tweedie, from Edinburgh, established himself in Buenos Aires as a plant hunter. Exploring the Sierra de Tan-

dil, 300 miles south of Buenos Aires, he found several new verbenas. The most important of these was the *Verbena platinius*, which, he reported, covered the mountain slopes in the area. He sent the seeds to the distinguished botanist William Jackson Hooker, and *V. platinius* was introduced to the British Isles through the botanical gardens at Glasgow, Liverpool, and Dublin. It eventually reached the United States, along with many minor species, most of which quickly spread as weeds to wastelands and roadsides.

Two latter-day English collectors, Clarence Elliot and H. F. Comber, scouring the Andes as late as 1928, discovered the *V. corymbose*, a hardy variety that extended the use of verbenas in many gardens. Its bushy plants and full, bright flower heads have made it today's most popular verbena.

Interest in the earlier verbenas had reached its peak in the late nineteenth century, when Victorian gardeners especially welcomed them and the plant hybridizers, who were just coming into their own, found their natural adaptability irresistible.

Creation of a Modern Plant

Our twentieth-century garden verbena is one of the loveliest and most satisfying products of man. A creation of the laboratory hybridists, it is a cross of the scarlet verbena of the Argentine, the purple verbena of Peru, and the white verbena indigenous to most of South America.

Under cultivation, these hybrids, which possess the tenacity of weeds and the brilliance of prima donnas, produce masses of large blooms from June to frost. The flower heads, two to three inches in diameter, are made up of twelve bright florets that contrast with masses of attractive dark-green leaves two to four inches long. They range in color through soft shades of blue, crimson, and white, and they may be solid or striped, or have contrasting or white eyes that betray their dominant ancestor.

One of the most popular species is the bush verbena, which grows eight to ten inches high and bears flat-topped brilliant flowers in endless profusion. Dwarf varieties of verbenas were also discovered in South America, and they, too, have been hybridized; today there are fifty varieties of dwarfs averaging from six to eight inches in height, which make ideally neat, compact borders in rock gardens.

The fern verbena, whose beauty is still little known, bears lavender and blue flowers and has delicate fernlike leaves. The sand verbena possesses a most subtle and exquisite odor, and its sprawling habit and delicate lavender flowers make it an excellent ground cover.

Since garden verbenas first appeared in American seed catalogues in 1890, the hybrid forms have had a phenomenal success. They are grown in gardens all over the Temperate Zone, thriving as a perennial in the south, and as a gratifyingly reliable annual in the north.

Suspect Relations

The verbena belongs to the Verbenaceae family which includes herbs, shrubs, and tender trees in various parts of the world. Verbenas are, as the old Spanish padres noted, close relatives of the European vervain, *Verbena officinalis,* a plant with a long history of participation in mystic and religious rites.

Centuries before the Christian era it was used by both Greeks and Romans in their worship of the gods. It was considered a sacred herb, and was used to cleanse the altars of Jupiter before a sacrificial feast. Through ages of such use, it became associated with other gods, and acquired a number of folk names: "Juno's tears," "herb of Venus," "Mercury's blood." Circe and Medea used it to cast spells on gods and men, and its properties are substantiated many times by Virgil, Horace, and Pliny the Elder, all men familiar with the activities of gods and all naturalists at heart.

The Roman legions carried vervain to their far-flung provinces, where its magic became part of the local pagan worship. In early Germany, villagers gathered about bonfires on Midsummer Eve, and cast garlands of vervain into the fires as a token sacrifice to the sun god, to entice him to return for another season. The Druids, too, revered the vervain's magical properties and used it in secret rites on Midsummer Eve. (See St.-John's-Wort.)

Later, though, the Crusaders returned from the Holy Land with seeds of vervain and tales of its growing on Mount Calvary, where it had been used by Christ's followers to stanch His blood and cleanse His wounds. New folk names sprang up and vervain was called the "herb-of-grace" and the "herb-of-the-cross," and was taken into monastery gardens, where it assumed a medicinal role.

By Elizabethan days it had become a cure-all recommended for everything from fever to falling hair, gallstones, and the black plague. The early herbalists Gerard, Turner, and Culpeper recommended it enthusiastically, and it became yet another of the herbs the Puritans carried to New England for their physic gardens. Josselyn, to whom we owe so much of our firsthand information on plants the early settlers grew, tells us that by 1673 it grew everywhere and healed everything. The settlers called it "simpler's joy," and used it for "sore throate, tooth ache, stomache gripe, and sword wounds."

Although Josselyn disparages such faith in vervain's universal efficacy and refers to it as "Clown's Heal-all," its use endured, and tradition says it was often used in the Revolution to stanch the blood of patriots' wounds.

Today vervain has escaped to the wastelands, roadsides, and wet meadows, its services and magic forgotten. In late summer it waves its tassels of rosy flowers, unrecognized, unwanted, and ignored, biding its time—a living clue to our colonial past. But to those who delight in the art of gardening, the modern hybrid verbena offers a roster of assets. It flowers abundantly in a splendid range of colors. It is adaptable to most soils. It can be used for bedding, edging, banks, rock gardens, and cutting, generating its own colorful magic in any summer garden.

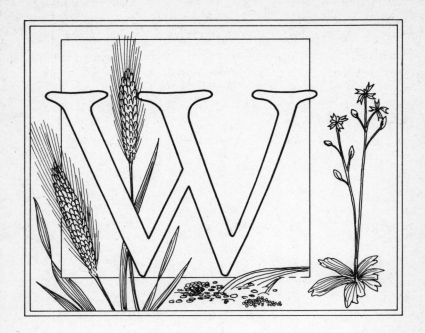

WATERCRESS

Watercress was introduced very early to the American colonies because it was a familiar and dependable plant that could be grown at once in the numerous fresh-water streams near the first settlements. It provided a salad green much of the year, or a boiled vegetable to serve with meat; more important, it was a powerful antiscorbutic, and scurvy was a real problem in early days of colonization.

Governor John Winthrop included cress in the first order of seeds from London for his Massachusetts garden, where it is listed as "Cress – 3d." We are not certain what variety of cress that was, but we may assume it was water-cress, whose planting would not have to wait for forests to be cut or garden beds laid out. This seems substantiated by John Josselyn's *New-England's Rarities Discovered*, published in 1672, which reports that watercress was already naturalized, thriving in America's streams, and commonly used by the colonists. And Leonard Meager, whose handbook *The English Gardener* was the authoritative guide for settlers for generations, lists watercress under essential greens for "Kichin and Sallet."

Watercress, like the early immigrants, quickly spread through the virgin land. Many old records, both private and public, show it was growing abundantly in June and July in fresh-water streams across the nation. According to John Kieran, the twentieth-century naturalist of New York, it even grew within the confines of the city.

History

Like flax and the crocus, watercress is one of the oldest plants associated with man. It is native to much of the temperate world and, from antiquity, has been used by sailors to relieve scurvy. In ancient Greece it was known as *kardamon* or head subduer. Xenophon recommended watercress to the Persians in 350 B.C., and Pliny the Elder cites it as a stimulant to the brain and a pleasant salad ingredient. The Romans, who loved it, carried it throughout their conquered empire and undoubtedly introduced it to England, where its Saxon name was *cerse*. It was cultivated at Saint Gall's monastery in Switzerland in 812, and grew in the physic gardens of many other monasteries as a beneficent herb. In powdered form it was an ingredient in a medieval ointment for sword wounds, whose formula came from the Basque country, where watercress was a universal food.

In Elizabethan England the herbalists Gerard and Culpeper wrote variously of watercress. Gerard's comments on the pleasure of the cresses in general extend to garden cress, winter cress, black cress, and Indian cress. Culpeper says that watercress is under the dominion of the moon, and that, beaten up with lard, it is an apt treatment for scales on children's heads. Other health notes from old books urge the daily consumption of watercress to increase a scholar's wits, to avoid measles, and as a relief from toothache if the juice of fresh cress is blown into the ear.

A more romantic note is a knight's pledge to his lady: "My love shall grow as the watercress / Slowly, but with deep roots."

Botany of Cress

Watercress is a perennial aquatic. It may grow entirely submerged or extend six to eight inches above water. In winter it may appear as a green in the bottom of a stream, and in spring spread with branching stems that root in the gravel bottom.

Watercress belongs to the crucifer family, like its close relatives the mustards, and bears small, white, four-petaled flowers that suggest a cross. Its botanical name, *Nasturtium officinale*, was given to it by Linnaeus. The Romans called it "nasturtium" or nose twister, though its taste is more pungent than its odor is sharp; *officinale* refers to its long service as a healing herb. It was known as "nasturtium" for untold centuries before the beautiful South American *T. majus* arrived in Europe. The plants are unrelated, but the similarity in flavor of the flowering nasturtium's aromatic leaves to the long-established watercress made it quickly acceptable as a salad green. In popular usage it soon usurped even the old Roman name, and has kept it ever since. (See Nasturtium.)

Cultivation Today

American streams of clean, cool, clear running water provided an ideal medium for the successful cultivation of watercress. In time the European variety, whose seeds the colonists had sowed, crossed with the native wild cress, and today this spontaneous hybrid can be found in streams from Nova Scotia to Georgia and westward to California.

Generations of European immigrants have used its small, pungent leaves as a salad green and condiment, but its considerable commercial value is limited by the fact that it deteriorates rapidly after picking. The water in which it grows must be entirely free of pollution—an increasing rarity in our reckless industrial world. But it is possible to grow it, too, in soil kept constantly very damp, in the shade. It is also grown extensively in greenhouses for winter markets. And there are city dwellers, insulated from the earth, who nostalgically grow watercress in their apartment pot gardens.

WEEPING WILLOW

The beautiful old tree stood near the door of an ancient saltbox house in a New England village. Its swaying branches drooped protectively over the path. Perhaps it was a descendant of America's first weeping willow; it grew near a stream not far from where that tree had been planted centuries ago.

How the first weeping willow came here is told by William Cullen Bryant, one of the most eminent of our early poets and for fifty years editor and part owner of the New York *Evening Post*. In a volume of his *Orations and Addresses*, Bryant gives credit to another distinguished early American, Samuel Johnson, an Anglican clergyman, philosopher, and author, who in 1754 was appointed the first president of King's College—which was to grow and prosper and become Columbia University after the Revolution.

Traveling in England, Johnson visited the house on the Thames in Twickenham where Alexander Pope had lived. On the grounds grew a giant weeping willow, a famous tree, which he admired so ardently that, when he returned to America, he took with him a gift of some cuttings from it. He planted them at his home in Stratford, Connecticut, on the

Housatonic River near Long Island Sound. The cuttings, which must have found the Housatonic as salubrious as the parent tree found the Thames, grew prodigiously; and in a few years Johnson was generously providing cuttings to the many visitors who came to see the graceful foreign tree, with its poetical associations. So weeping willows began to spread along the eastern waterways.

Bryant believed that these trees were probably the progenitors of all the weeping willows in the region, where they grew rapidly to a size that he had never seen them attain in any other part of the world.

Weeping Willows in the South

The earliest printed record of the weeping willow in America is somewhat more recent: an advertisement in the *Maryland Gazette*, dated 1790, offered one that had just arrived in Baltimore with a shipment of plants from an English nursery. Once introduced, weeping willows rapidly became popular as ornamentals throughout the south—partly because of their association with famous persons and places.

George Washington was among the first to grow them. Benjamin Latrobe, considered to be the first professional American architect, visited Mount Vernon in 1796, and made this note: "A level lawn, bounded on each side with a wide but extremely formal serpentine walk shaded by weeping willows." They must have been very young at the time, for Richard Parkinson, an English writer who traveled in America from 1798 to 1800, says that when he saw them he was struck with surprise, and that Washington told him the trees had been introduced to America only recently.

Weeping willows also grew on James Madison's estate, Montpelier. Dolly Madison writes of them in her memoirs, in 1809 or thereabouts, in terms that suggest they were still considered a rarity. Travel notes by other English visitors comment on the weeping willows used in the landscaping of many great plantations of the south. Brandon, the estate in tidewater Virginia that had been established by a friend of Captain John Smith, seems to have had numerous weeping willows.

Napoleon and the Weeping Willow

Hudson Lowe, the governor of Saint Helena when Napoleon was a prisoner there, had found when he first arrived that the volcanic island was destitute of trees and flowers. Being an Englishman and a gardener, he asked to have some trees and flowers sent out to make his island more homelike. Among the shipments was a weeping willow, which he ordered planted beside a spring. It responded to the mild climate, and grew quickly, as healthy willows do, making a pleasant shady spot. A garden seat was placed there, and Napoleon went to it so often that the tree was called "Napoleon's willow."

Shortly after his death a windstorm destroyed the tree, but cuttings from it were planted about his grave, and grew into a small grove—a living memorial.

Before the Suez Canal was opened, Saint Helena was a stop on the long ship routes around the Cape of Good Hope. For years the passengers who visited Napoleon's tomb carried weeping-willow cuttings away, and planted them in many parts of the world.

Darwin, as he relates in his *The Voyage of the Beagle*, visited Napoleon's grave when his ship anchored off Saint Helena in May 1836. As he explored the island and learned its history, his admiration for Hudson Lowe grew.

When the Portuguese discovered Saint Helena in 1502, it was covered with great ancient trees. But a garrison of the British East India Company took it over in 1651 and imported large flocks of hogs and goats to raise for food. Roaming at large, the animals destroyed all the ground cover; they ate the native flowers and the young trees, and the old trees were felled and used for fuel. Two centuries later the destruction was irretrievable; scarcely a tree remained on the island, and any other native plants had receded to the highest peaks. The volcanic island had become desolate indeed. But by Lowe's orders the animals were restricted, and some green grass returned. Then began the importation of flowers and trees from Australia and England, and they must have prospered. Darwin counted 746 different plants on the island, of which only fifty-two were indigenous. He notes that weeping willows were now common along the riverbanks, and that Scotch firs and blackberry hedges gave the island a pleasant English air.

The Weeping Willow in England

But the weeping willow was not English. It had arrived in England in the late 1500's and must have been known to Shakespeare; at least it would seem appropriate that the willow in Desdemona's song and the willow aslant the brook on which Ophelia hung her garlands were weeping willows.

A story told about the first weeping willow in England concerns a Mr. Vernon, an English merchant who fell in love with a weeping willow tree he found in Aleppo on the banks of the Euphrates River. He took a cutting of it back to England with him, and planted it on his estate at Twickenham, on the Thames. Here it grew to a great age, and was a great curiosity to Englishmen. There is little doubt that Alexander Pope's willow began as a cutting from this tree, which may therefore have been the actual source of the original American willows. Through the years it drew so many crowds that in 1801—or so it is said—the man who then owned it cut down the patriarchal tree in the hope of enjoying a quiet and peaceful privacy.

China and the Willow Pattern

Linnaeus classified the weeping willow as *Salix babylonica*, but later research has established that its original home was neither Babylon nor Turkey, but southern China, where it is inseparably bound up with old legends, and with antique embroidery and art.

Yet willow-pattern dishes originated not in China but in Staffordshire, England. It is true that the design itself, created about 1780 by Thomas Minton, an apprentice potter, is said to be based on an old Chinese legend. It portrays the garden of a mandarin, and a pair of lovers attempting to elope across a bridge beneath the concealing branches of a weeping willow, with "Three birds flying high / Two boats sailing nigh."

The ware became enormously popular, and in time the design was copied by potters in many parts of the world—including the Orient, where willow-pattern ware is still manufactured in quantity, and exported chiefly to the West.

Botanical Bits

All willows belong to the world-wide family of deciduous trees and shrubs called Salicaceae. They are members of the *Salix* branch, and all of them are easily propagated by cuttings. Many are native to North America, among them the black willow of the south, the swamp willow of the Great Lakes area, the long-leafed willow of the Midwest, and the pussy willow, that early harbinger of spring.

The willows are easily identified by their narrow alternate leaves, soft light wood, bitter bark, and the male and female catkins that the trees bear in spring. The willow is said to have derived its botanical name from the ancient Celtic words *sal*, near, and *lis*, water.

The weeping willow is still one of the most popular trees for landscaping. It also grows wild along thousands of streams across America. Few dream that it did not always do so.

WHEAT

There was no wheat in America when the white man came, nor was there any rye, or oats, barley, or millet. The only grass the Indians grew for food was corn, *Zea mays*.

America's wheat harvests may now reach a billion bushels annually. How did this miracle happen?

How Wheat Began in America

In 1493 Columbus planted wheat in the West Indies "to prove the soil," and reported to their Spanish majesties that it flourished. This opened up for Spain a dazzling prospect: a new source of wheat to augment her own limited harvests, and perhaps eventual control of the European wheat market through imports from the New World.

There was reason for such confidence. From the Moors, who had occupied the southern part of the country for centuries, and who had introduced highly sophisticated techniques, Spain had inherited the most advanced agriculture in Europe. As early as the twelfth century the Moors had produced a handbook on the production of grains and fruit, and its wisdom had a far-reaching influence on early Spanish American colonization, and on the settlement of our own southwest.

Spain's dream seemed near reality in 1519, when Cortez told of the advanced agriculture of the Aztecs he had conquered. He had found in Mexico the world's first botanic garden—acres of land devoted to the cultivation of plants collected from far places—and also an agricultural college to which men came from every part of the Aztec kingdom.

In 1529, when Spain's occupation of Mexico was secure, the conquistadors were given tracts of land, and carpenters and bricklayers were dispatched to build commodious haciendas for them. King Ferdinand ordered that a shipload of women be sent to Mexico, too, to ensure a dominant Spanish population. Finally, in 1529, a shipload of wheat arrived, for planting fields that would be cultivated by the Indian slaves and establish a Spanish wheat empire in the New World.

But the Indians had never seen wheat and knew nothing of its cultivation. Completely different from maize, it took three times as long to grow and produced less food. The Spanish colonists had also become dependent on maize, and few of them were trained in farming. The result was limited production of wheat on the great estates.

Nevertheless the project went forward, under the auspices of Mother Church. The center of every pioneer village in Spanish America was a mission with a garden, an orchard, and grain fields, and the padres taught the Indians European agriculture along with the Catholic faith.

By 1687 wheat was being widely grown in the viceroyalty of New Spain. The Jesuit missionary Father Eusebio Francisco Kino, who founded a series of missions in what is now northern Sonora and southern Arizona, introduced wheat to the Indians of the region. When Father Junípero Serra established a chain of missions in California (see Orange; Rose), wheat, brought by ships from the West Indies, was planted in the mission fields. By 1782 3,000 Indians had been converted and were cultivating wheat fields and orchards. Presidios, or military posts, were established to protect these flourishing missions, and the settlement of California was ensured.

But Spain's estate system, dependent on unskilled, enforced labor, and her mission agriculture, dependent on a few paternal monks and reluctant natives, belonged to the Middle Ages. All this fell before the vigor of the new United States, which offered free land to free men with the ambitious dream of possessing their own farms.

Wheat in the Colonies

Almost all American colonists brought wheat with them from Europe. It was growing at Jamestown in 1611 and at Plymouth in 1621, and Governor William Bradford reported that wheat bread was eaten at the first Thanksgiving. It was planted by the Swedes on the Delaware and the Dutch on the Hudson. It grew and thrived with the skill of the farmer, the quality of the soil, and the beneficence of the climate. During the Revolution, Vermont's wheat production gave the state the title "Bread Basket of America."

As our country expanded beyond the Appalachians, wheat went with the pioneers. Tennessee and Kentucky celebrated their first important wheat harvest in 1777.

Corn went with the settlers, too. It was raised to feed their cattle, hogs, horses, sheep, and chickens, and the great barns built to house the abundance were the expression of a dream fulfilled. For the first time in the history of the world, meat came to be accepted as an essential daily food in a nation's diet.

While America grew, Europe was torn by wars that interfered with agriculture and urbanized millions of people in manufacturing centers, where hunger and unrest grew when bread became scarce. England, France, and Germany lifted their tariffs on wheat; they sent their ships to America; they rivaled one another in the price they offered for America's grain.

But much of America's wheat was still sown by hand and cut by scythe or sickle. There were not enough men to harvest it, or enough horses and wagons to transport it, or enough river boats to take it to ports. And so, once again, necessity became the mother of invention.

Technology for a Wheat Empire

In 1837 John Deere perfected the steel plow, and in 1843 formed a partnership for its manufacture with Leonard

Andus. In 1842 two Pennsylvania farmers, the Pennock brothers, dug up a design for a seed drill that Jethro Tull, an English agriculturist and inventor, had tried out more than a century before, without much success. From Tull's drill the Pennocks developed a machine whose modern version can plow, pulverize, and fertilize the soil, drill rows, and plant seed mechanically.

In 1847 Cyrus Hall McCormick arrived in Chicago, a pioneer town on the edge of the wheat country, and began the full-scale manufacture of a mechanical reaper he had first successfully demonstrated in 1831, and patented three years later. It was a dramatic moment in world history, because millions of Europeans were waiting for bread, and American wheat was spoiling for want of harvesters.

Constantly improved reapers streamed from Chicago factories, and in a few decades the city was transformed. In 1840 her population had been less than 5,000. By 1870 it had risen to 300,000. Great grain elevators of steel and concrete, built near water and rail centers, rose overnight, and Chicago became the world center for wheat exchange.

Forests and prairies were falling before the axes and plows of eager pioneers, and immigrant wheat was replacing the prairie grass. But a new and greater problem arose: how to get the wheat harvests to markets and ports.

Wheat and Railroad Building

In 1840 our nation had only a few thousand miles of railroad tracks, and most railroads were local ventures, designed for transporting people. Suddenly "freight for wheat" became a national cry, and "build railroads" was the answer. By 1880, 156,000 miles of rails had been laid, and wheat traversed the nation to waiting ships that carried it to Europe's hungry.

Europeans who were looking for safe investments outside their nations, where the masses were revolution-prone, eagerly contributed to the tremendous financing involved in our rapid development of railroads. America boomed. Steel plants opened in Pittsburgh and along the Ohio to manufacture rails, locomotive works designed new engines, and boxcar plants opened in eastern cities.

Wheat and Canals

But wheat hadn't waited for rails and trains. Some wheat farmers had looked to rivers as access to ports—rivers made navigable and extended by canals. In 1825 the construction of the Erie Canal was completed, linking the Hudson River with Lake Erie. This, and continuing navigational improvements to the Ohio River, which linked Pittsburgh with the Mississippi, changed the status of the Midwestern farmer and inspired canal building along many eastern waterways. The abundance of wheat that his rich lands produced had brought only twenty-five cents a bushel in a local market, and a barrel of milled flour only $1.25. When the grain boats carried his flour to the cities in the east, the Middle West, and the south that could be reached by water, it sold for $8.00 a barrel.

Industrialized Europe became dependent on American wheat, and millions of European peasants dreamed of American abundance. Hundreds of thousands of immigrants began pouring into our Atlantic ports—some 40 million of them in the first 150 years of our nation. The demand for wheat grew, and our government stepped in to increase wheat production.

Wheat and the Department of Agriculture

The Department of Agriculture decided we must make maximum use of our northwestern states for growing wheat. In 1898 the department sent Mark Carleton to Russia to find a spring wheat to complement the winter wheat that was planted in Kansas and the southwest in the fall and harvested the next summer. Such wheat cannot survive the deep winters and spring frosts of northern states.

Carleton, who had experimented with hundreds of varieties of wheat and who knew the Russian language, was well prepared for his assignment. The result of his Russian visit was the introduction of durum wheat to the United States. This is the hardest-kerneled of all wheat, rust-resistant, and adapted to dry conditions, and it brought a whole new dimension to American wheat culture. Two years after its introduction it produced 60,000 bushels; and by 1907, 45 million.

Wheat Mills

Wheat was one of the basic factors considered when the first colonists decided where to locate their towns. It was essential to build near a waterfall or stream with enough power to turn a mill wheel to grind their grain. Every settlement had its mill, and the farmers carried their grain to the miller and took home flour for bread.

Today's mills, no longer run by water power, are highly mechanized, and besides grinding finer flour they sift, grade, and package it. The bran—the coarse outer coat—is separated and packaged for cattle feed. Whole-wheat flour, which deteriorates more rapidly than white because of its content of wheat germ and bran, is specially labeled. Durum wheat is milled for macaroni and spaghetti manufacturers. Soft winter wheats are milled into the finer flour used for cakes, pastry, and biscuits.

Thirty percent of most wheat harvests goes into the manufacture of important by-products of the mills, such as wallpaper paste, glue, adhesives for plywood, starches, and textile sizing. The mills ship the greater part of the flour to commercial bakers.

Mass production of bread by commercial bakeries was originated by W. B. Ward in New York in 1849. With the slogan "Untouched by Human Hands," Ward built a billion-dollar business. Modern mechanized bakeries increased the demand for wheat, because wheat makes the best bread.

Wheat flour is rich in gluten, which makes the dough strong and elastic, capable of expanding when the yeast is added, and the result is a light, raised bread. The only other grain with a comparable gluten content is rye, but in America, where the production of wheat far outstrips rye, rye bread remains on the whole a specialty. (See Amaranth.)

Increased production of wheat has also changed the bread of Europe in the last few centuries. In 1700, 40% of the English ate black rye bread; today less than 1% do. In 1700 Poland and Russia used three times as much rye as wheat, but by now they have reversed this statistic and use three times as much wheat as rye.

History Briefs

The history of wheat, a member of the grass family, genus *Triticum*, goes back thousands of years, but its origin is a mystery. Twentieth-century research suggests that it first appeared in the Nile valley, where it was probably cultivated as early as 5000 B.C. By 4000 B.C. it was growing in the Indus and Euphrates valleys, by 2500 B.C. in China, by 2000 B.C. in Europe west to England.

Wheat has run through the legends and religions of the Mediterranean world since the dawn of history. Its name in English appears to have two derivations, the Sanskrit *sveta* and the Old English *hwaete*. Both words mean "white," and suggest how wheat flour differs from the dark flour of other grains used for bread.

Modern wheat has been hybridized to adapt to many climates, and experimentation continues in all wheat-growing countries in a constant effort to improve existing varieties, or to develop new ones with greater disease resistance or hardiness or productive capacity. So widespread is its cultivation that not a month passes when it isn't being harvested somewhere. The United States, China, and Russia are the largest producers of wheat, which is also grown extensively in India,

Western Europe, Canada, Argentina, and Australia. The Ohio valley, the prairie states, and eastern Oregon and Washington are the chief wheat-growing areas in our own country, with Kansas leading in production. Each year the harvest starts in the south and moves north. The great reapers pass over the land, cutting the golden grain, an impersonal, mechanical, efficient process. But the cry of the machine is still the prayer of the man who reaped with the scythe: "Will there be bread for all?"

XANTHIUM

Xanthium is a word with magic in it, conjuring up visions of glamorous flowers and mysterious gardens in the Far East. But in prosaic fact it is the botanical name of the common cocklebur, a familiar weed of our fields and roadsides, and in the Old World a curse to farmers and shepherds since Biblical times.

The cocklebur, according to an early edition of Asa Gray's *Manual of Botany*, was given this curious name by Joseph Tournefort, a French botanist who investigated the flora of Asia Minor from 1700 to 1702 and originated the botanical classification of plants that was used until Linnaeus devised our present system. It was the custom of the time to commemorate famous men and historical places by naming plants after them. One theory holds that Tournefort chose *Xanthium* as the generic name for the cockleburs to commemorate Xanthus, the ancient capital of Lycia, which had been a thorn in the flesh to Alexander the Great when he was building his empire in Asia Minor. A modern and more plausible explanation is that *xanthos* is the Greek word for yellow, and one of the characteristics of the cockleburs is their thick yellow sap, which was long used in Greece as a hair dye.

The European cocklebur arrived in America as a stowaway with the first loads of sheep and goats brought here by the early settlers. It has spread from coast to coast and is now familiar to every farmer, shepherd, and hiker—and to their dogs.

Two forms of this immigrant plant are prevalent in the eastern states: *Xanthium strumarium,* which has heart-shaped leaves and oblong, hairy burs with strong, beaked prickles, and *X. canadense,* which may have been brought to America by the earliest French settlers along the Saint Lawrence. By now it has spread throughout the Great Lakes region and along the rivers and creeks of the Middle West, and is often thought to be native. The most familiar variety of this cocklebur, which has oblong burs and incurved, beaked prickles, is found in all the warm parts of the world.

Another variety of *X. strumarium* appears to have arrived in America with the colonists from Germany who settled in Pennsylvania. There it is called "clotbur," a corruption of *Klette,* German for stick and bur; its botanical name is *X. pensylvanicum.* It dominates the fields of all the north central sections of our country.

While these European cockleburs were invading the north, another species, *X. spinosum,* a native of tropical America, was moving in from the south and spreading along our East and West coasts and the Gulf area, accidentally introduced by the early Spanish settlers. It is known as "thorny clotbur" after the sharp beak possessed by each of its burs, which are only a third of an inch long. It is easily distinguished by its hairy stems, and by yellow spines at the base of its lanced-shaped leaves.

A more decorative species of the tropical American cocklebur is the Texas star, X. *texarium,* which once covered hundreds of acres of the dry, open prairies. American ranchers now keep it under control, but many gardeners of the southwest cultivate the Texas star, whose bright-yellow solitary flowers, two to three inches in diameter, provide color where poor soil and intense heat preclude more delicate plants.

Cockleburs and Burdocks

Today cockleburs are often confused with burdocks, another European immigrant weed now common in America. The burdocks are biennials, with round, many-seeded, hooked burs. The great burdock, *Arctium lappa,* has been traditionally used as a medicinal in Eurasia, and even today is cultivated as a vegetable in Japan.

Botanical Bits

All cockleburs or clotburs belong to the Compositae family, as do the burdocks. The Xanthiums, however, are annuals, from one to three feet high, with stout branching stems and alternate, toothed leaves, generally heart-shaped. Sterile and fertile flowers appear on various heads of the same plant. The seeds enclosed within the rough, prickly burs are two-celled, with an ovule in each cell. These ovules germinate in consecutive seasons—a design of nature that ensures their self-propagation and defies man's attempts to exterminate them.

In South Africa cockleburs became a national problem when European settlers established their great sheep ranches. The young plants poison the livestock that eat them, and the burs reduce the quality of the wool, so the government passed strict laws for the eradication of this threat to the national economy.

In ancient times, the cocklebur was considered a beneficent herb. The roots, seeds, and leaves of the Eurasian weed, *X. strumarium*, were used as a tonic to purify the blood, and as a diuretic. The leaves of the tropical American cocklebur, *X. spinosum*, were used by the Indians to counteract hydrophobia. The transitory life of the cockleburs, like that of many weeds, exemplifies the ingenuity with which plants both serve and challenge humankind.

"The Yarrow :
As disagreeable
among flowers
as a cynic is
among men."
Frank Bolles
1891

Achillea millefolium

YARROW

Old records show that man was using yarrow a thousand years before the birth of Christ. Today it is a roadside weed, its ferny leaves and asymmetrical white flowers recognized by few, ignored by most.

But in Greek mythology the soldiers under Achilles carried yarrow with them when they stormed the walls of Troy. Achilles had been a pupil of the centaur Chiron, who was famed for his medical knowledge, and who taught him that yarrow leaves possessed the property of stanching blood from the wounds of arrow or spear. So we may think, if we like, that yarrow helped the Greeks to conquer Troy. From the legend, it takes its botanical name, *Achillea millefolium*.

The Roman soldiers of actual history carried yarrow leaves to stanch blood and sterilize wounds, and the barbarian tribes who invaded Rome's Mediterranean empire carried it home with them to central Europe. By the eighth century Britons were using it, and it appears in Saxon leech books of the year 1000 under the Latin name *millefolium* or "thousand leaves." But *millefolium* soon took on the Anglo-Saxon name *gearwe*, of which our modern word "yarrow" is a corruption. It has also carried the folk name "staunch weed," in recognition of its medical properties.

During the Middle Ages, yarrow grew in every monastery and castle herb garden. It was often made into ointments and salves with beef and mutton tallow; steeped in water or wine, its leaves produced bitter potions used for various ailments, or they were made into a tea like that brewed from its close relative camomile. But whereas camomile tea was most commonly used as a tonic, yarrow tea was taken to control melancholy, and even became suspect as a witch's brew. In the Scottish Highlands, yarrow was long used to reduce fevers and control falling hair, and Gerard pronounces that "yarrow leaves chewed, especially greene, are a remedie for toothache."

Yarrow in America

Yarrow was among the herbs the early colonists brought to America, where it continues to serve as a dependable blood coagulant. It was also used in colonial times, before the importation of hops, to clarify and flavor beer. Linnaeus observed that beer thus treated was much more potent than beer flavored with hops. The Indians appropriated yarrow and grew it about their villages, using it not only as a coagulant but also for reducing swollen tissues, and like Gerard's Elizabethans, for earaches and toothaches.

The Shakers planted it in America's first commercial herb garden at Lebanon, New York, where they held it in high respect, and raised, packaged, and sold it for generations. It was carried west in many covered wagons and planted in the gardens of many isolated pioneer cabins. Today it grows in almost every state, a common denizen of fields and wastelands from the Atlantic to the Pacific.

Modern Service

Norman Taylor, a twentieth-century authority on plants, says there is no scientific evidence for yarrow's medical properties, but contemporary horticulturists have found a new use for it. Planted in a vegetable garden, as experiments have shown, it not only acts as a pest repellent and an agent against blight, but it also seems to encourage the production of vegetables of superior size and quality.

In the flower garden, too, yarrow appears in modern guise. From among the hundred or more species that grow in the Temperate Zones, nurserymen have produced a number of delightful hybrids. Many of them bloom from June into September. They are available in pure white, various shades of yellow, and rose red, and there are dwarf varieties ideally suited to rock gardens.

But yarrow's special gift for modern gardens is its subtle fragrance—a pungent, spicy odor that it retains when it is dried and made into winter bouquets.

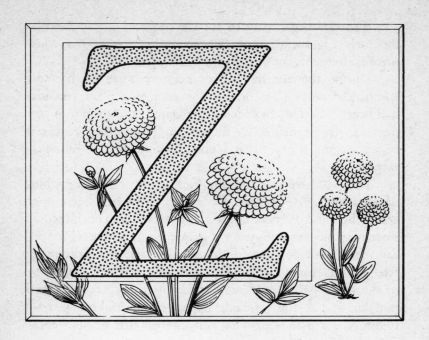

ZINNIA

The sophisticated zinnia that we grow in our gardens is a very modern flower, whose story is a long pastoral with a brilliant twentieth-century ending.

It began in 1519, when the conquistadors marched through Mexico, tramping over a low-growing, insignificant weed with small, dingy-purple or dull-yellow blossoms that seemed to spring up everywhere. In the surfeit of truly fabulous Aztec plants, the sturdy annual that grew so prolifically and bore infinite numbers of small daisylike flowers completely escaped attention. The Aztec name for it meant "eyesore," which in Spanish became *mal de ojos*, and the Spanish colonists in Mexico continued to tramp over it for two centuries more.

The tide began to turn for the neglected, ugly weed in the early eighteenth century, when there arose the widespread, intense interest in all things botanical.

During the first half of the century, most of the more spectacular, appealing, or readily obtainable plant treasures had been collected, classified, propagated, and grown in gardens far from their place of origin. Now plant explorers began seeking flowers that were still unknown, if less spectacular.

In 1750 there was a young German doctor in the school of medicine at Göttingen University who had become an ardent botanist. Pioneering in wildflowers was considered an extremely elegant hobby for a gentleman, and Dr. Gottfried Zinn had made a name for himself by doing so. When his friend the German ambassador to Mexico sent him some seeds of the weed *mal de ojos* he was delighted. Spectacular it was not, but it had the romantic appeal of the unknown, so it briefly aroused the curiosity and interest that greeted every new plant from the Americas.

Young Dr. Zinn died in 1759, and the new flower was named "zinnia" in his honor. But another century passed before it attracted much attention. It was grown in a few European gardens as a curiosity, seeming to thrive where nothing else would—which inspired such common names as "everybody's flower," "poorhouse flower," and "garden Cinderella." But that was the extent of its fame, and it was unknown in the gardens of the United States.

In the 1880's some French horticulturists began experimenting with zinnias. Their first breakthrough was with a dwarf form, possibly not unlike the modern dwarf perennial zinnias, which are thrifty and adaptable and bear a mound of small dull-colored flowers with repressed petals. The new flowers were startling enough to encourage further experimentation. In 1886 France produced a double zinnia in bright clear colors with a larger flower than any previously grown. New interest was awakened, and nurserymen in the United

States took zinnias under consideration. Here are the vital statistics they had to work with:

⊸§ There were, they discovered, sixteen species of zinnia, all indigenous to Mexico. The improved zinnias of Europe were apparently derived from the Mexican *Zinnia elegans*, and the dwarf form, Z. *haageana*.

⊸§ Because zinnias are short-day flowers from the sub-tropics, full sunlight is essential to them.

⊸§ The zinnia's weedy history made it tolerant to any type of soil or climate, though it flourished best in deep rich loam, and in warm weather.

⊸§ Its weedy nature also gave it sturdy stem growth, and clasping opposite leaves protected by bristly hairs and a slightly pungent odor discouraging to insects.

⊸§ The pigments in the original dull-colored flowers had already been synthesized to produce a range of purples, reds, pinks, yellows, and whites. There were no blues.

⊸§ The flower heads were solitary, made up of a limited number of ray flowers and a dense head of disk flowers similar to those of the aster and daisy families of the order of Compositae.

This final characteristic was most important to the hybridizers. They transformed the scrawny single wildflower by forcing the tiny center disk flowers to assume the character and shape of ray flowers, and thus created the showy, brilliantly colored modern zinnia.

Hybrid Zinnias

The history of modern hybrid zinnias originated with Luther Burbank three-quarters of a century ago in California. By 1920 he had produced a dahlialike zinnia, a breakthrough that would lead to years of exciting experiments, and to today's flowers, which range from giants to dwarfs, with single and double and crested blossoms, and brilliant solid and blended colors.

After Burbank died his head gardener, William Henderson, carried on the zinnia research. William Atlee Burpee took over after he bought the Henderson Seed Company in the early forties. But he found that his famous Floradale Farm, in Lompoc, California, was too near the Pacific fogs for this native of sunny Mexico, and so he purchased land in Santa Paula, ninety miles to the south and twelve miles inland, especially for experiments with zinnias.

Millions of zinnias were planted over the years, because only 5% pass the annual selection, based on color and character, of those retained for further research. Moreover, the former Mexican weed proved stubborn in some respects, and refused complete surrender of her individuality.

Tetraploid Transformation

Nature has designed a special pattern for every plant in the world. As research reveals these patterns, hybridizers have set about improving them.

Every cell in a plant or an animal contains a fixed number of chromosomes—a chromosome being the structural carrier of inherited characteristics. The average plant is a diploid, with two sets of chromosomes in the nucleus of its cells. (See Tiger Lily.) When it is disturbed by freezing, extreme heat, or some other natural cause, it may change its whole character, and the number of chromosomes in its cells may double. Such plants are called tetraploids, and if their seeds prove fertile they may produce plants with larger and more complex flowers.

Burpee understood the tremendous commercial possibilities of artificially shocking the zinnia into becoming a brilliant tetraploid flower, one susceptible to unlimited but controllable variety, and began an arduous program for hybridizing zinnias. By 1948 the hybrids were still only average tetraploids. Their colors were brilliant, but they were unmarketable because their seeds could not be depended upon to produce plants that would repeat the parental characteristics. The zinnia still resisted. Burpee still insisted.

Then, in that very year, one plant in a huge field of zinnias turned out to be wholly female. Its flower heads looked like pincushions filled with pinlike stigmas. They had no petals, no pollen-laden stamens, but the plant could be bred with Burpee's best male-flowering zinnias. Eventually it made possible the first generation of hybrid zinnias on a commercial scale.

This freak zinnia was discovered in the experimental fields at Santa Paula, in Row 66. In time it came to be called "Old 66," and its strong characteristics were successfully used in further hybridizing. After ten more years of experiment and selection, the F1 or first-generation hybrid zinnias were firmly established, went on the market, and became one of the wonders of the flower world. Old 66 was the mother of them all.

More wonders followed, and today's catalogues offer chrysanthemum, dahlia, and button zinnias; cactus, giant, and ruffled zinnias; cupid, pompon, and mini zinnias; and, finally, the lovely cut-and-come-again zinnias that simply ask to be cut to bloom all season long. Some blossoms measure seven inches across, others no more than one; some plants grow to two and a half feet, others to no more than a decorous six inches. A flower bed planted wholly to zinnias could be marvelously varied.

As a result of this triumphant horticultural experimentation, the modern zinnia ranks second among all garden flowers, surpassed in popularity only by its cousin the marigold. It is the state flower of Indiana, and it has advocates as ardent as the marigold's for being chosen our national flower as well.

Like so many green immigrants, the *mal de ojos,* the eyesore over which the conquistadors heedlessly trampled, has come a long way to make our gardens beautiful.

INDEX

CLAIRE SHAVER HAUGHTON was born in Fairmount Springs, a village in Pennsylvania her ancestors founded before the Revolution. A botany major in college, she found her interest gradually focusing on the romance of immigrant plants and the part they had played in American history. The lack of source material was a challenge, and research into this whole aspect of botany became a dominant interest in a life that included, besides marriage and three children, a career of teaching and incidental writing. Mrs. Haughton read her way through mountains of books; visited botanical gardens in Mexico, Hawaii, and Europe as well as in the United States; and, down the years, filled many notebooks with her discoveries. So *Green Immigrants* was born.

Mrs. Haughton and her husband now reside in a big old house in Wallingford, Connecticut.